Tarkeshwar Barua, Ruchi Doshi, Kamal Kant Hiran
Mobile Applications Development

Also of interest

Analog and Hybrid Computer Programming
Bernd Ulmann, 2020
ISBN 978-3-11-066207-8, e-ISBN (PDF) 978-3-11-066220-7
e-ISBN (EPUB) 978-3-11-066224-5

C++ Programming
Yuan Dong and Fang Yang, 2019
ISBN 978-3-11-046943-1, e-ISBN (PDF) 978-3-11-047197-7
e-ISBN (EPUB) 978-3-11-047066-6

Elementary Synchronous Programming.
in C++ and Java via algorithms
Ali S. Janfada, 2019
ISBN 978-3-11-061549-4, e-ISBN (PDF) 978-3-11-061648-4
e-ISBN (EPUB) 978-3-11-061673-6

Computational Methods for Data Analysis
Yeliz Karaca and Carlo Cattani, 2019
ISBN 978-3-11-049635-2, e-ISBN (PDF) 978-3-11-049636-9
e-ISBN (EPUB) 978-3-11-049360-3

Web Applications with Javascript or Java.
Volume 1: Constraint Validation, Enumerations, Special Datatypes
Gerd Wagner and Mircea Diaconescu, 2018
ISBN 978-3-11-049993-3, e-ISBN (PDF) 978-3-11-049995-7
e-ISBN (EPUB) 978-3-11-049724-3

Tarkeshwar Barua, Ruchi Doshi,
Kamal Kant Hiran

Mobile Applications Development

—

With Python in Kivy Framework

DE GRUYTER

Authors

Tarkeshwar Barua
Department of IT
BlueCrest University College
Monrovia, Liberia, West Africa
tbarua1@gmail.com

Kamal Kant Hiran
School of Engineering
Sir Padampat Singhania University
Udaipur, Rajasthan, India
kamal.hiran@spsu.ac.in;
kamalhiran@gmail.com

Ruchi Doshi
Department of Computer Science and
Engineering
Jyoti Vidyapeeth Women's University
Jaipur, Rajasthan, India
doshiruchi18@ieee.org

ISBN 978-3-11-068938-9
e-ISBN (PDF) 978-3-11-068948-8
e-ISBN (EPUB) 978-3-11-068952-5

Library of Congress Control Number: 2020950010

Bibliographic information published by the Deutsche Nationalbibliothek
The Deutsche Nationalbibliothek lists this publication in the Deutsche Nationalbibliografie;
detailed bibliographic data are available on the Internet at http://dnb.dnb.de.

© 2021 Walter de Gruyter GmbH, Berlin/Boston
Cover image: scyther5/iStock/Getty Images Plus
Typesetting: Integra Software Services Pvt. Ltd.
Printing and binding: CPI books GmbH, Leck

www.degruyter.com

Preface

Mobile phones will change the way we connect to the world in future. Mobile phones are becoming a very important part of daily life in today's digital age. It is time-consuming for programmers to develop mobile applications for various mobile platforms such as Windows, iOS, Linux, and Raspberry Pi. There are numerous platforms available in the market, such as Android, iOS, Windows, Linux, and Sun, and each platform is different due to internal architecture, which is why all platform of software is different. Cross-platform mobile application development using python with the Kivy framework is the best approach that works well on all mobile application platforms. The Kivy framework is easy to understand and user-friendly. Nowadays, the software development industry is becoming very competitive and the consumer wants their high-quality application with multiple device support in a short span of time; therefore, the cross-platform application development market is going to be very high. These apps make life much easier by offering remote access to a wide range of services.

In this book, we cover the fundamentals of the python in the first part, which is required to build any mobile application, and in the second part, the principles of mobile application development using the Kivy framework. The second part discusses the responsive graphical user interface designing, customization, various file operations such as sending/receiving and reading/writing over local/remote machines, and network operation over remote resources. All chapters are written based on the real-time examples and practical knowledge. Exporting required executable file according to the platforms along with packaging and licensing. Each chapter has summary, key terms, review questions, and exercise for better learning. At the end of the book, some hands-on projects will also help readers to strengthen their practical and programming skills.

Chapter 1 presents an overview and characteristics of mobile application, difficulties during development process, advantages and disadvantage of Python, installation of Python on various available platforms such as Windows, Linux, Mac, and machine compatibility issues due to different platforms and architectures.

Chapter 2 presents the variables such as declaration, data types, operators, string handling, functions, and math calculations.

Chapter 3 presents the conditional statements, loops, continue and break statements, array, string, lists, tuples, set, and dictionaries.

Chapter 4 gives an overview of the object-oriented programming concepts: abstraction, class and object, polymorphism, inheritance, encapsulation, overloading, overriding, access specifier, and decorators.

Chapter 5 describes the standard libraries, file operations using native, exception handling, import standard Python modules and packages, regular expression, JavaScript Object Notation, SQLite3, and SQLAlchemy.

https://doi.org/10.1515/9783110689488-202

Chapter 6 is an introductory chapter of Kivy framework, which presents the cross-platform, Kivy architecture, installation of Kivy framework on various platforms, and anatomy of Kivy.

Chapter 7 presents the basics about Kivy, how to configure working environment on various platforms, Kivy framework configuration and how to import Kivy library, widgets, orientation horizontal and vertical, super, padding, __init__, and add_widget.

Chapter 8 provides several types of layouts such as box layout, float layout, anchor layout, grid layout, relative layout, stack layout, and page layout with examples.

Chapter 9 presents the designing user interface (UI) that is targeted to meet the responsiveness. It describes the UI components, buttons, labels, ListView, recycler view, events and properties, popup menus, TextInput, ScrollView, and KivyClock.

Chapter 10 presents the UX widgets: actionbar and slider, checkbox and text on window, toggle button, TreeView, radio/checkbox button and label management, ProgressBar and carousel layout, custom ProgressBar, scatter layout, canvas layout, AsyncImage, spinner, accordion, and switching between two screens.

Chapter 11 presents graphics handling, modules, network support, audio, and video animations.

Chapter 12 describes the packaging apps for various platforms: mixin class, building an android APK on OS X, application testing with Android and iOS, packaging application using Kivy launcher, and Kivy application testing.

Chapter 13 introduces two sample hands-on-practice projects.

The book targets the academic and programming needs of the mobile application development community. It would be used by students and faculty members interested in the fields of computer science and engineering, and information science and technology engineering. The authors would be more than pleased if the readers found it useful to discuss further ideas in the development direction of cross-platform mobile applications.

Tarkeshwar Barua

October 2020   Dr. Ruchi Doshi

Udaipur, India   Kamal Kant Hiran

Contents

Authors' Biography

Tarkeshwar Barua has completed MCA and MBA (IT) from Punjab Technical University Jalandhar, India. He has more than 10 years of experience in information technology (IT) industry as developer and as corporate as well as academic trainer. He has hands-on working experience in developing various stand-alone and web-based application packages. He is working as senior lecturer at the BlueCrest University College Liberia and Senior Full Stack Developer (Python and Java) Vodafone India. Earlier, he worked as technical corporate international trainer in EY Dhaka for leveraging information and communication technology. He has worked as programmer in companies such as AV Systems (New Delhi), 360 Degree Info Dynamics Pvt Ltd (Noida), and Symbiotic Pvt Ltd (Chennai). He has organized various national and international corporate classroom/online training and conferences. His area of interest is in mobile app development, machine learning, and deep learning using python and R language. He has become a leading resource in the field of IT consulting and development.

Dr. Ruchi Doshi is having more than 12 years of academic, research, and software development experience in Asia, Europe, and Africa. Currently, she is working as a research professor at the Azteca University, Mexico, and adjunct professor at the Jyoti Vidyapeeth Women's University, Jaipur, Rajasthan, India. She worked in the BlueCrest University College, Liberia, West Africa, as registrar and associate professor; BlueCrest University College, Ghana, Africa; Amity University, Rajasthan, India; Trimax IT Infrastructure & Services, Udaipur, India. Dr. Ruchi is the founder chair of the RKB Research Foundation based in the lake city Udaipur, Rajasthan, India. She has several awards to his credit such as international travel grant for Germany from ITS Europe (Passau, Germany), IEEE Liberia Subsection WIE Founder Award, IEEE Senior Member Recognition, IEEE Student Branch Award, Elsevier Reviewer Recognition Award, and the Best Research Paper Award from the University of Gondar, Ethiopia. She has published 20 research papers in peer-reviewed international journals SCI/Scopus/Web of Science and conferences; 2 Indian patents; 3 books on cloud computing, mobile cloud computing, and intelligent Internet of things (IoT) systems for big data analysis. She is a reviewer and editorial board member of various reputed international journals in Elsevier, Springer, IEEE, Bentham Science, and IGI Global. She is an active member in organizing many international seminars, workshops, and conferences. She has done many international visits in Denmark, Sweden, Germany, Ghana, Liberia, Ethiopia, and Jordan for research exposures.

Kamal Kant Hiran works as an assistant professor at the School of Engineering, Sir Padampat Singhania University (SPSU), Udaipur, Rajasthan, India, as well as a research fellow at the Aalborg University, Copenhagen, Denmark. He is a gold medalist in M.Tech. (Hons.). He has 15+ years of experience as academician and researcher in Asia, Africa, and Europe. He is recipient of several awards and recognition from IEEE and Elsevier. He has published 35 research papers in SCI/Scopus/Web of Science journals, conferences, 2 patents, and 7 books in International Publishers. His research interests focus on cloud computing, intelligent IoT, and framework development by using ubiquitous computing.

https://doi.org/10.1515/9783110689488-204

1 Introduction of Python

> Python is an experiment in how much freedom programmers need. Too much freedom and nobody can read another's code; too little and expressiveness is endangered. ~Guido Van Rossum

1.1 Why Python

Python is a very popular and widely used general-purpose interpreted programming language. It features a dynamic type of system and has automatic memory management with multiple programming paradigms including imperative, functional, and procedural. Programs can be created to automate various tasks. Nowadays, all major IT companies such as Google, Microsoft, and Apple, are using Python as the primary programming language. Python is the most accessible programming language to be learnt quickly in a short period. It can develop any application such as desktop, web application, mobile application, and utility tools, with the help of a wide range of frameworks and libraries available with minimum coding. It is free and open-source available with proper documentation over the Internet. Python focuses on business logic instead of the basic facts of programming. It is available in two versions 2.x and 3.x but primarily, the focus will be on 3.7 and the upcoming Python 3.8. Python 3.8 will be full of new features with its main emphasis on performance-related issues. Python is dynamically typed, automatically collects garbage, and manages memory with multiple programming paradigms such as procedural, object-oriented programming (OOP), and functional. Python is an agile language that allows the user to do many customized things within a short span of time. The application development can commence from any point of the module and, in the ending phase of application development, can join each other just before releasing.

1.1.1 Drawbacks of Python

All programming languages have some advantages and disadvantages. Python also has some disadvantages; some of them are listed as follows:
1) Python is very slow in terms of execution of the program in comparison to other programming languages such as C, C++, and Java. Java is the fastest language because of Java Virtual Machine (JVM) and Just In Time (JIT) compiler. For further information, one can refer to YouTube videos.
2) Drawing sophisticated graphics is computationally quite expensive. This reduces graphics quality.
3) Without Cython code, execution is very slow. Cython enables us to compile our code at C level. C compiler optimizes the things.

https://doi.org/10.1515/9783110689488-001

4) GPU or Cython is required for better performance.
5) We can limit the speed of the application in order to use vast resources.

1.2 History of Python

Python is not named after the type of snake. It is named after a British comedy group called "Monty Python's Flying Circus." Python language was created by "Guido van Rossum" in 1991 as 1.0. After 10 years, the 2.0 version was released in the year 2000 with new features like comprehensions, cycle-detecting garbage collection, and Unicode support. Rossum was a big fan of the group and their quirky humor. Python programs often use the group jokes and famous quotes in their code as a tribute. With various major updates, Python is available in two major versions Python 3.x and 2.x. Python 2.x is a legacy version that would continue up to the year 2020 while python 3.x is a fast updated and popular version. Some features can be imported from python 3 using __future__ in Python 2.x. Python 3.0 was released in the year 2008, and it was the major release without backward compatibility, which means Python 2.x code could not run over the 3.x series. This major issue was fixed in python 3.7 version for a large amount of code had been written in 2.X series. Initially, this support was offered up to 2015 but now it has been extended up to 2020. Rossum solely took this responsibility up to July 2018, and now he shares his leadership as a member of a five-person steering council. Now, the steering council will release the next upcoming versions.

1.3 Major features of Python

- Easy to learn and read because of the tabular space instead of curly braces.
- Using the English language, python program becomes very easy for coding.
- Python is based on Interpreted language, which saves a lot of time for the execution of the application.
- Procedural programming, OOP, and functional programming support; using thread module Python supports multi-threading.
- Python Android and iOS templates parse to ease generating a dynamic page for the client.
- Interactive language; Python code never converts human-readable code for execution. This feature makes Python very interactive by making changes at run time.
- Python supports Android and iOS by KIVY framework.
- Platform-independent; same code can be executed on all available platforms.

1.3.1 What is new in upcoming Python 3.8

Assignment expression is known as *walrus operator* (:=). It allows value to be assigned to a variable, even to a variable that does not exist. It follows list comprehension and lambda function tradition:

```
input_text=input("Please input text> ")
while input_text!="quit":
    print("You have entered ", input_text)
    input_text=input("Please input data> ")
```

and the above-given program can be written as

```
while (input_text := input("> ")) != "quit":
    print("You entered:", input_text)
```

Positional only parameters help us to remove ambiguity, as which argument is to be in a function definition position and which are to be keyword arguments. We can achieve it with the help of forward-slash (/) and star (*), where (/) indicates that some function parameters must be specified prepositionally and cannot be used as a keyword argument. We can define positional only parameters for functions. Here, developers can force some arguments to be positional only. Observe the given example:

```
>>def myfunction (a,b,/,c,d,e,*,f,g,h):
>>. . .  print(a,b,c,d,e,f,g,h)
>>myfunction(1,2,3,4,5,6,7,8)
Traceback (most recent call last):F
    File "<pyshell#15>", line 1, in <mmyfunction(1,2,3,4,5,6,7,8)
    TypeError: myfunction() takes 6 positional arguments but 8 were given
in the given example is applicable after python 3.8 and above
```

Python initialization and configuration Python 3.8 adds a new C API to configure the Python initialization providing better control on the whole configuration and better error reporting. Python code configuration is scattered all around the code. It helps to read and modify configuration before it is applied.

Vectorcall is the fastest calling protocol for Cpython. It is the default implementation of Python, and it compiles the source code into intermediate byte code, which is executed by the Cpython virtual machine. It allows for far faster internal Python methods without the overhead of creating temporary objects to handle the call. Any extension-type implementation is callable, and we can use this protocol.

Run time audit hooks provide two APIs in the Python run time for hooking events and making them observable to outside tools like testing frameworks or logging and auditing systems. An audit hook is an exit point that allows the auditor to add the modules subsequently. This happens by activating the hook to transfer control to an audit module. The verified open hook allows Python administrators and embedders to integrate with the operating system support while launching scripts or importing python code. An audit hook and a verified open hook allow applications and frameworks written in pure Python code to take advantage of extra notifications.

Pickle protocol 5 without band data buffers helps us to transfer compatible data separately from the main pickle stream, at the discretion of the communication layer. The process in which a Python object turns into byte stream is known as "pickling" or "serialization" or "marshaling." When pickle is used to transfer large data between the Python processes in order to take full advantage of multicore, it is important to optimize the transfer by reducing memory copy.

Typing-related hints Python 3.8 is more mature now with the release of some new features such as literal types, typed dictionaries, final objects, and protocols. Typing defines a standard notation for Python function and variable-type notations. The notation can be used for documenting code in a concise, standard format, and it has been designed to also be used by static and run time-type checker, static analyzer, Integrated Development Environment (IDE), and other tools. Python supports type hints, in the given example number should be a float and double() function should return a float, as well. Python treats this annotation as hints. They are not enforced at run time. Here acceptFloat() function accepts string too as an argument.

```
def acceptFloat(number: float) -> float:
    return 2 * number

print(acceptFloat(2.45343))
print("This is not a float value")
============================output=========================================
4.90686
This is not a float value
===========================================================================
```

There are several static type checkers available like Pyright, Pytype, Pyre, and MyPy that come along with PyPi.

Parallel file system cache for compiled byte code Python 3.8 introduces a new PYTHONPYCACHEPEFIX setting that helps to configure the implicit byte code cache to use a separate parallel file system. Cache is reported in **sys.pycache_prefix** and **none**.

Debug build shares application binary interface (ABI) as release builds Python 3.8 uses the same ABI whether it is built in release or debug mode. Now it is possible to load C extension built using stable ABI. Python 3.8 has release build and debug build, both are ABI compatible. On Unix, C extensions are no longer linked to libpython, except on Android and Cygwin. It is also possible to statically link Python to old C extensions using a shared Python library.

f-string supports a handy (=) specifiers for debugging It is a form of debugging where print statements are inserted to print the value of expressions or variables that needs to be tracked with wring logs. Logs are very useful in the production environment.

f-string was introduced in python 3.6. We can evaluate an expression as part of the string along with inserting the result of function calls.

```python
name="Tarkeshwar Barua"
print(f"Hello, Mr. {name}")
print(f"Hello, Mr. {name.capitalize()}")
print(f"Hello, Mr. {name.upper()}")
print(f"Hello, Mr. {name.lower()}")
print(f"{name.upper() = :-^10}")
print(f"{name.upper() = :->10}")
print(f"name={name}")
print(f"{name=}")

==============================output====================================
Hello, Mr. Tarkeshwar Barua
Hello, Mr. Tarkeshwar barua
Hello, Mr. TARKESHWAR BARUA
Hello, Mr. tarkeshwar barua
name.upper() = TARKESHWAR BARUA
name.upper() = TARKESHWAR BARUA
name=Tarkeshwar Barua
name='Tarkeshwar Barua'
========================================================================
```

continue is now legal in finally block A continue statement cannot exit a finally block when a continue statement occurs within a finally block. The target of the continue statement must be within the finally block, otherwise a compile time error occurs. The older version of python does not allow continue in a finally block because all code after the continue will never be executed. It was there, otherwise, its interpretation would have been problematic.

The default asyncio event loop is now ProactorEventLoop in Windows The asyncio stands for asynchronous input–output and refers to a programming paradigm that achieves high concurrency using a single thread of event loop. The event loop is the place where most of the magic happens in asyncio. The asyncio module, introduced in Python 3.4, provides convenient shortcuts to access the methods of the global and default policy. Tornado requires the add_reader family of methods, so it is not compatible with the ProactorEventLoop on Windows. It is currently providing two implementations of event loops SelectorEventLoop and ProactorEventLoop. The ProactorEventLoop is incompatible with SSL and the default loop supports SSL on Windows. By extension, aiohttp should also work with https on Windows. The event loop is based on the selector's module. It supports Subclass of AbstractEventLoop to use the most efficient selector available on windows sockets; Proactor event loop for windows uses IOCP (I/O Completion Ports).

The spawn start method is now used by default in multiprocessing (Mac) The multithreading tests were failing on Linux for everything except Python 3.4. Spawn is a Python package that allows users to concisely specify and execute many tasks with complex and codependent input parameter variances. It is used where thousands of similar simulations with input parameter variations are run for design and certification purposes. It allows the specification of such large sets to be formulated in a concise input file.

```python
from multiprocessing import Pool
from os import getpid

def calculateSquare(i):
print("Process No. ", getpid())
return i ** 2

if __name__ == '__main__':
with Pool() as pool:
result = pool.map(calculateSquare, [1, 2, 3, 4, 5, 6, 7, 8, 9, 10, 11, 12, 13, 14, 15])
print(result)

================================Output==================================
Process No. 21044
Process No. 21456
Process No. 21044
Process No. 21456
Process No. 21044
Process No. 21456
Process No. 21044
```

```
Process No. 21044
Process No. 21456
Process No. 5180
Process No. 21456
Process No. 5180
Process No. 21044
Process No. 21456
Process No. 5180
[1, 4, 9, 16, 25, 36, 49, 64, 81, 100, 121, 144, 169, 196, 225]
========================================================================
```

Multiprocessing can now use the shared memory segment to avoid pickling cost between process allows regions of memory to be created and shared between different Python processes. In earlier versions of Python, data could be shared between processes only by writing it out to a file, sending it over a network socket, or serializing it using the pickle module. Shared memory provides a faster path for passing data between processes, allowing python to more efficiently use multiple processors and processors core. Shared memory segments can be allocated as raw regions of bytes or they can use immutable list-like objects that store a small subset of Python objects – numeric types, strings, byte objects, and the None object.

Type_ast is merged back to Cpython There is a fork of the ast module (in C language) named *typed_ast* used by mypy, pytype and IIRC also by some linters. Its redeeming quality preserves certain comments. It is very hard to keep this code up to date with developments in the grammar of the language.

By-default now pickle uses protocol 5 to improve performance provides a new way to pickle object that implements Python buffer protocol such as byte, memory views, and NumPy arrays. Pickle is used to transfer data between Python processes in order to take advantage of multicore or multimachine processing, it is important to optimize the transfer by reducing memory copies and possibly by applying data-dependent compression. Pickle cuts down on the number of memory copies that must be made for such objects.

The above-mentioned information is given based on the published papers by https://docs.python.org/3/

1.4 Market demand

There are several programming languages being used to develop application in the market, but I would like to introduce some facts that are taken from various sources of the Internet. These facts are given as primary introductory language, sector-wise demand, salary wise, and market share.

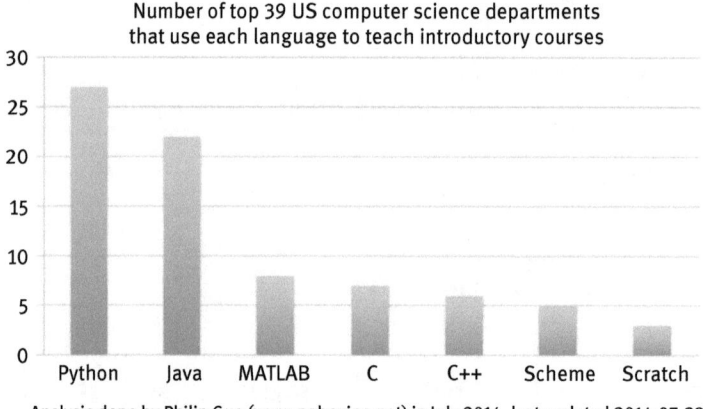

Number of top 39 US computer science departments
that use each language to teach introductory courses

Analysis done by Philip Guo (www.pgbovine.net) in July 2014, last updated 2014-07-29

Figure 1.1: Market demand for programming language.

2016 Average Developer Salary in the U.S.

indeed.com estimations(USD)	Language
# 1 115,000	Swift
# 2 107,000	Python
	Ruby
# 3 104,000	C++
# 5 102,000	Java
	C
# 6 99,000	JavaScript
# 7 94,000	C
# 8 92,000	SQL
# 9 89,000	PHP

Figure 1.2: Language-wise average salary.

Worldwide, Nov 2016 compared to a year ago:

Rank	Change	Language	Share	Trend
1		Java	23.4%	−0.5%
2		Python	13.7%	+2.4%
3		PHP	9.8%	−0.9%
4		C#	8.4%	−0.4%
5	↑↑	Javascript	7.6%	+0.6%
6	↓	C++	7.1%	−0.7%
7	↓	C	7.0%	−0.5%
8		Objective-C	4.7%	−0.5%
9	↑	R	3.2%	+0.5%
10	↓	Swift	3.2%	+0.3%
11		Matlab	2.6%	−0.1%
12		Ruby	2.0%	−0.3%
13	↑	VBA	1.5%	+0.1%
14	↓	Visual Basic	1.4%	−0.5%

Figure 1.3: Language-wise popularity.

1.5 Why Python in mobile app development?

Currently, mobile phones have become a significant part of our daily life; therefore, it is almost impossible for people to live without their mobile phones. Mobile phones have made our life very easy. Keeping this in mind, today most of the software companies have moved to mobile app development. Yet, there are a few challenges to develop a mobile app due to the constraints of various mobile phone platforms such as Android, iOS, and Windows, and all these platforms have different software requirements. It means programmers need to write code for each and every platform natively, which is a very time-consuming job; therefore, cross-platform is a good approach. Python is very easy and rich in library support that makes app development very easy. Figure 1.4 illustrates the market demand in various sectors, which proves the importance of Python. Python is considered easy to learn and runs almost anywhere. It provides strong support for integration with several technologies and higher programming productivity across the development life cycle. It is good for large and complex projects with changing requirements. It is

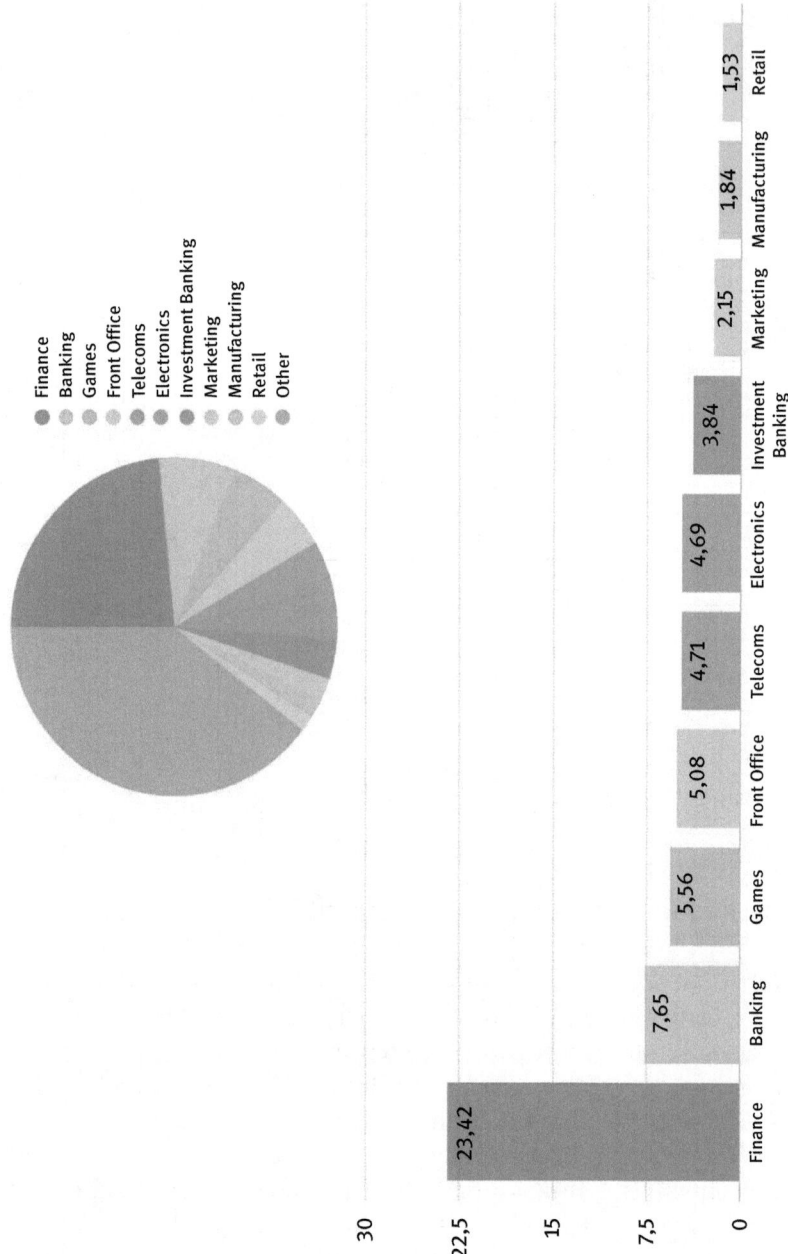

Figure 1.4: Industry sector-wise market demand.

the fastest growing language and runs on several millions of phones including various industries. Python codes are easy to read.

1.6 Python versions

Table 1.1: Python versions.

Version	Release date
Python 1.0	January 1994
Python 2.0	October 2000
Python 2.1	April 2001
Python 2.2	December 2001
Python 2.3	July 2003
Python 2.4	November 2004
Python 2.5	September 2006
Python 2.6	October 2008
Python 2.7	July 2010
Python 3.0	December 2008
Python 3.1	June 2009
Python 3.2	Feb 2011
Python 3.3	Sep 2012
Python 3.4	March 2014
Python 3.5	September 2015
Python 3.6	December 2016
Python 3.7	June 2018
Python 3.8	October 2019
Python 3.9	October 2019

We can check the installed Python version using the following command at terminal/command prompt:

```
$ python --version
        Or
$ python3 --version
```

1.7 Architecture of python application

Python syntax and style is flexible when it comes to structuring the application. There are many opinions about its structure, but here we have given the most commonly used structure by the programmers. It is very useful for managing and distributing application easily. A basic program structure is given herein, where "*helloworld*" is the folder name/project name that contains all the required file:

gitignore – It tells git which file needs to be ignored, like configuration files, IDE clutter.

helloworld.py – This is the main program to execute which contains all the Python code.

LICENSE – It is a plain text file that describes the license of the project. It is mandatory to have a license while distributing code. This file name is always in all caps letter.

README.md – Stands for Markdown file (project documentation file). It is very important to craft this file; we can create this file easily by the URL https://dbader. org/blog/write-a-great-readme-for-your-github-project.

requirements.txt – Here we can specify all our program dependencies along with their versions. Adding dependency in this file called pinning, pip freeze command can be used to add all the production dependency in this file. It is useful only in a developing environment. Adding dependency into this file is called pinning. Pip and freeze commands are used to do so:

```
pip install -r < requirements.txt
```

setup.py – This is similar to what the requirements.txt file does but in package redistribution. This file is useful in end-user system dependency installation, where user indicates site-packages directory for running Python:

```
python setup.py install --user
```

test.py – This file is responsible for the tests, to ensure proper working of the application.

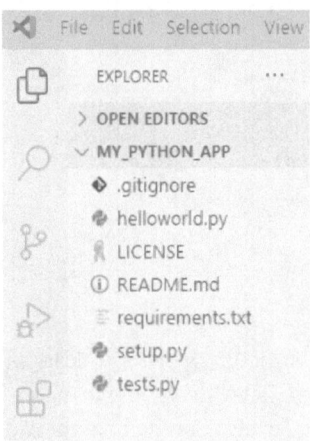

Figure 1.5: A Python program structure.

1.8 System requirement

All the given codes in this book are valid for both Python versions, but for the best result, tested environment is with Python 3.x as given in Table 1.2:

Table 1.2: System requirement.

Specification	Description
Python version	2.x (2.3.1 to 2.7.16) and 3.x (3.3.0 to 3.7.3)
Processor	Pentium-4 or higher
RAM	512 MB (1 GB recommended)
Operating system	Windows XP (incl. 64 Bit) or higher, Linux (any), Mac OS X 10.5, Raspberry Pi,
Screen resolution	800*600

1.9 Installation of python in various platform

Installation in windows

After downloading Python **.exe** package installer from https://www.python.org/, double-click on it and follow the instructions as given, click next, and finally finish. We need to set windows environment path to temporary, which can be set from the command line. The command is given as:

```
C:\Users\admin>set path= "C:\Program Files\python\bin\"
```

and permanent path can be set using environment variable by right clicking on My Computer → Properties → Advanced → Environmental Variable → System Variables → Path → Edit → add you python path after putting; (semi-colon), Then OK,

Installation in Linux

In Linux, Python installation may differ according to various distributions but for most popular distribution Ubuntu, installation command is given as:

```
sudo apt install python
               Or
sudo apt install python3
```

for other distribution of Linux we can download Gzipped source tar file from the https://www.python.org/ and install it from Terminal and add Python executable location to the PATH environment variable. Python can be installed from the source code too; we need to download it from the Internet. After unzipping source code we need to apply ./configure command and finally make command to generate executable add environment PATH:

```
pip install -r < requirements.txt #if you have already installed python
```

```
wget --no-check-certificate https://www.python.org/ftp/python/2.7.X/
Python-2.7.X.tgz tar -xzf python-2.7.X.tgz
cd Python-2.7.X
./configure
make
sudo make install
```

Installation in Mac

Mac comes with preinstalled python, but it can be outdated. We can download Python **.pkg** package from https://www.python.org/. We double click on it and follow the instructions as given and finally finish, and add Python executable location has been added to the PATH environment variable.

1.10 Creating First Hello World app

We can write Python script in any ASCII/ANSI-based editor such as IDLE, Eclipse, Netbeans, Visual Studio, PyCharm, Notepad, Notepad++, Sublime, and Jupiter Notebook. It requires to download and install Python interpreter according to the operating system from https://www.python.org/:

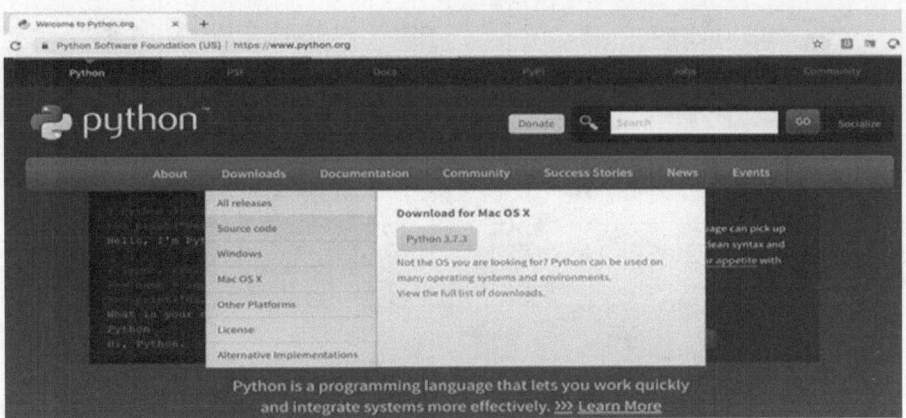

Figure 1.6: https://www.python.org home page.

After downloading and installation, we can write code built-in text editor IDLE, and snapshot is given later.

```
● ○ ●                         Python 3.7.2 Shell
Python 3.7.2 (v3.7.2:9a3ffc0492, Dec 24 2018, 02:44:43)
[Clang 6.0 (clang-600.0.57)] on darwin
Type "help", "copyright", "credits" or "license()" for more information.
>>> print("WelCome To Python")
WelCome To Python
>>> |
```

Figure 1.7: Python command line interpreter.

Writing program in python is very simple, as we have seen in the given snapshot. Here ≫ symbol indicates the line where we need to write our code/instruction, and this code can be interpreted when we press enter button and it will display the result immediately. These commands are sequential. We can save these codes in our file with extension name **.py** as given in the snapshot. We can close/exit from the python shell using exit/quit command as suggested here:

```
quit()            #ctrl+D
         Or
exit()            #ctrl+C
         Or
   # command+q in mac
```

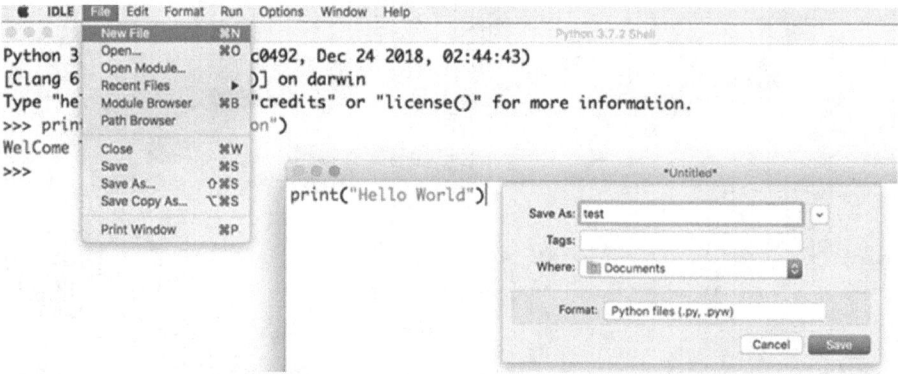

Figure 1.8: Writing and saving code using IDLE editor.

We can execute this code by just pressing F5 button in IDLE or using command python test123.py using command line/Terminal, as given later in the snapshot.

```
-bash: export: '/anaconda3/etc/profile.d/conda.sh': not a valid identifier
(base) admins-MacBook-Air-7:~ admin$ cd Documents/
(base) admins-MacBook-Air-7:Documents admin$ python test123.py
Hello World
(base) admins-MacBook-Air-7:Documents admin$
```

Figure 1.9: Python code execution on the command line.

Hurray! We are done with first Python program successfully.

These codes can be written as an arbitrary:

```
$ python -c 'print("Hello World")'
$ python -c 'print("My name : {0}, city : {1}, phone : {2}".format("Mr
ABC","Monrovia", "34453534534"))'
$ python
>>>print("Hello, ", end= "\t")
>>>print("World")
>>>print("Hello, ", end= "\n")
>>>print("Hello, ", end= "<br>")
>>>print("Hello, ", end= "BREAK")
>>>import sys
>>>sys.stdout.write("Hello World")
```

1.10.1 Basic operations in python prompt

We can perform basic operation on python prompt such as addition, subtraction, multiplication, and division as follows:

```
>>>10*3
30
>>20-15
5
>>>4+5
9
>>>5/2
2.5
>>>5%2
1
>>>5-10
-5
>>>10**3
1000
>>>12//8
1
>>12%8+(12//8)*8
12
>>>10--2
12
>>>a=10
>>>a+=2
>>>a
12
```

1.11 Anatomy of Python program

Python programs are easy to write, but it becomes difficult while we develop a large application, which involves many files and codes. It becomes more difficult to manage when it goes into errors; hence, some standards are suggested to avoid this type of difficulties. Some of the rules are:

1. How one program is connected to other programs?
2. Code should be written in the isolated form (various small units), so that in case of errors, it can be fixed within short period of time.
3. Do not repeat yourself (DRY), never writing same coderepeatedly, or just reading it.

4. OOP can make our program more manageable such as classes, methods, and objects.
5. Special configuration and log files must be located at some location for input/output.
6. Always writing comments in the program about its functionality to understand flow of program.

1.12 IDLE

Integrated Development and Learning Environment (IDLE) is an alternative to the command line, as the name suggests, it is very useful for developing new code for learning python. While other interpreters can be used to execute python program. The default IDE comes with python interpreter can be used for its various advantages, some of which are listed as follows:
1) Multiwindow text editor with syntax highlighting, auto-completion, and smart indent
2) Automatic indentation
3) Integrated debugger with stepping, persisting breakpoints and call stack visibility
4) Python shell with syntax highlighting
5) Saving the python program as **.py** files and run them and edit them later at any them using IDLE
6) Python program can be run using **F5** key python shell to run interpreter

1.13 User input

To get input from the user, we use the input function (raw_input if we are using python 2.x). This function takes string as an argument, which displays the prompt and returns a string and waits until user provides input. By default, it takes string, but it may type caste automatically by input data and then it may require converting it into required data type:

```
>>> name=input("What is your name?")
```

1.14 Installing external modules using pip

When it comes to installing external modules there are several ways but most efficient is pip. First, we must make sure pip is installed in our machine, any way it comes with Python. Using the following command, we can check pip version:

```
$ pip --version          #for python 2
$ pip3 -version          #for python 3
```

For instance, with both Python 2 and Python 3 installed, pip often refers to Python-2 and pip3 to Python-3, both are different. If we are using python-2 then we use pip and for Python 3, we use pip3. We can do various tasks using pip, some of them are mentioned as follows:

```
$ pip search <query>                    #to search package
$ pip install [package_name]             #to install latest version of package
$ pip install [package_name]=x.x.x       #to install specific version
$ pip install '[package-name]>=x.x.x'    #to install minimum version package
$ pip -proxy http://<server_address>:<port> install [package_name] #while
you are behind proxy server
$ pip list -outdated                     #list of all outdated packages
$ pip install [package_name] -upgrade     # to upgrade a package
$ pip install -U pip # to upgrade pip version
$ python -m pip install -U pip
$ py -m pip install -U pip
```

Here x.x.x stands for version number of packages which we want to install. There is no utility of upgrading all at one time, we need to upgrade packages one after one.

Summary

- Guido Van Rossum has developed python language; it was needed because of the complex design theories of all other programming languages.
- Python is very popular as it is easy to learn and dynamically type caste language.
- Most of the programming languages are considered as difficult because of the complex object-oriented mechanism, but python supports developing any program without object-oriented language. Python internally implements OOP automatically.
- Python is a general-purpose language. Market demand in various perpectives like salary, number of projects developed in various programming language, and current market share of all available programming language, for python both are high.
- Python is in huge demand in various sectors that is why we should consider Python for the development of cross-platform mobile applications.

- Python comes with various major releases with major changes yearly, like the architecture of python application along with its all-important component and along with their usages.
- Python is available in all the existing major operating systems. Keeping this in mind, the author has described step-by-step Python installation in the various available platforms such as Windows, Linux, and Mac.
- Beginning of any programming language is a major step to write the first program, we can create and execute our first python application "Hello World."
- Being a general-purpose programming language, it can perform basic arithmetic operations in python scripting. Anatomy of python application is given.
- Any text editor can be used to write Python. The program must be written in UTF-8 encoding. python editor by default comes with a python interpreter as a default IDLE, where we can write Python programs.
- Programmers can accept input from the user with the **input** functions. Further, these inputs can be used to process information.
- Any programming language cannot provide all the features. We can install third-party libraries using **pip** command. Pip command is used to manage packages in the development, testing, and production environment.

Key terms

Python programs are easy to write, but it becomes difficult when we develop a large application. Its effective design/architecture plays a very important role in dependencies, compatibility, and distributed environment: downloading and installing a Python version with respective system requirements; Python installation in various operating systems and environment variable settings; and creating and executing a hello world program using IDLE. One of the various advantages of IDLE is that with several sample codes it likes to get input from the user. In the production environment, it requires specific/latest versions of libraries, and pip is the most used tool for module/dependency/library download.

Review questions

1. Open browser and go to http://python.org and download Python for your operating system and install it along with its documentation.
2. From the python prompt, get help from the given terminal, use help() command prompt, and the check all the available documentation.

3. Practice basic calculator operations such as addition, subtraction, multiplication, division, and modulo on python prompt.
4. If you are driving a car at 45 km/h, how long will it take to reach 10 km distance?

Exercise

Tick the correct option

Q.1. Python supports multiple programming paradigms like _____
   a) procedural
   b) OOP
   c) functional programming
   d) all the above

Q.2. Python is very slow because _____ is/are absent.
   a) JVM
   b) JIT
   c) Compiler
   d) all the above

Q.3. Which key is used to run/execute the python program?
   a) F5
   b) F6
   c) F7
   d) F8

Q.4. Python's official Editor is _____.
   a) PyCharm
   b) Netbeans
   c) IDLE
   d) Eclipse

Q.5. Python was released in _____
   a) 1990
   b) 1991
   c) 1989
   d) 1992

Q.6.   What is pinning? pip freeze command can be used for _____.
a) adding dependency in Readme file
b) adding dependency in py file
c) adding dependency in xml file
d) adding dependency in requirements.txt

Q.7.   Which of the following commands is used in all production dependencies in one file?
a) ls, dir
b) pip, freeze
c) cd, mkdir
d) python, pip

Q.8.   DRY stands for _____
a) do not repeat yourself
b) design reference year
c) defensive rushing yards
d) dependent responding Y-axis

Q.9.   Who invented Python _____
a) Dennis Ritchie
b) Charles Babbage
c) Guido Van Rossum
d) Sundar Pichai

Q.10. Which command is used to download python packages or libraries _____
a) pip
b) git
c) pit
d) jit

Answers

Q.1.   d) all of the above
Q.2.   d) all of the above
Q.3.   a) F5
Q.4.   c) IDLE
Q.5.   b) 1991
Q.6.   d) adding dependency in requirements.txt
Q.7.   b) pip, freeze

Q.8.   a) do not repeat yourself
Q.9.   c) Guido Van Rossum
Q.10. a) pip

Fill in the blanks

1. Python language is named after a British comedy group called _____.
2. Python is the easiest programming language to _____ in _____ of time.
3. Python basically focuses on _____ instead of basic facts of programming.
4. Some features can be imported from python 3 using _____ in python 2.x.
5. _____ is the most efficient way to install external libraries.

Answers

1. Monty Python's Flying Circus
2. learn quickly, short period
3. business logic
4. __future__
5. pip

2 Python basics

> A Language that does not affect the way you think about programming is not worth knowing.
> ~Alan J. Perlis

2.1 Variables

A Variable is an identification name that is given to the memory location (memory location can be found by using id(x) method), let us say we want to store value 5 in variable x, where x is the name of memory location where value 5 has been stored and "=" is assignment operator a is used to cope reference from one variable to another variable. A variable is a name that refers to a value, whenever it is required, we can retrieve it by just calling its name x and it will provide a value using the given syntax:

<variable name> = <value>

Example:

```
>>>x=5
>>>id(x)
4518782192
```

In the above-mentioned example, it will store value 5 at some memory location and this memory location is assigned with three variable names, it is just like one person has three different names. A value is one of the basic things a program works with like a letter or a number, = symbol is used to assign value to any variable; assigning a value to a variable itself declares and initializes the variable with that value. No keyword can be declared as a variable because those have some special meaning in python language. A variable cannot be declared without assigning it a value. This assignment process goes from left to right, which means a variable like (5=x) is declared as invalid:

```
>>> x=y=z=5              #assigning a value to a variable
>>> id(x)                #print memory location of listed variable
1891092464
>>> id(y)
1891092464
>>> id(z)
1891092464
>>>5=x                   #Wrong variable declaration
```

https://doi.org/10.1515/9783110689488 002

```
>>>tarkesh_123_barua= "Author Name" #valid variable
>>>name= "Mr. TB"
>>>Name= "Mr. TB"              #name and Name variables are different
>>>x=y=z=5               #id(x) id(y) id(z) 4518782192
>>>a, b, c, d=10, 20, 30, 40     #different values can assigned
>>>print(abc)                 #variable must be defined before use
Traceback (most recent call last):
File "<stdin>", line 1 in <module>
NameError: name 'abc' is not defined
```

If we change its value, then it will create new memory location, otherwise all three variables will share the same memory location. Let us have a look at the snapshot:

```
Python 3.7.1 (default, Dec 14 2018, 13:28:58)
[Clang 4.0.1 (tags/RELEASE_401/final)] :: Anaconda, Inc. on darwin
Type "help", "copyright", "credits" or "license" for more information.
>>> x=5
>>> id(x)
4518782192
>>> x=y=z=5
>>> id(x)
4518782192
>>> id(y)
4518782192
>>> id(z)
4518782192
>>> x is y
True
>>> x=y
>>> x==y
True
```

Figure 2.1: Variable declaration.

Keywords – Keywords are the special characters that have some special meaning for programming language. These are reserved words. Some of them are mentioned in Table 2.1

Table 2.1: Keywords.

False	True	class	finally	is	return	None	continue	for	lambda
try	def	from	import	nonlocal	while	for	and	del	global
not	with	as	if	else	elif	or	yield	assert	pass
break	except	in	raise	async	await				

Source: https://www.programiz.com/python-programming/keyword-list

There are certain rules for declaration of a variable:

1) Variable name must start with a character and underscore.
2) Variable can contain letters, numbers, and underscore.
3) period(.), hyphen(-), and special characters (`,~,!,@,#,$,%,^,&,*) are not allowed.
4) Keywords cannot be a variable name.
5) Variable names are case-sensitive (hello, Hello, HELLO, hELLO are different).
6) No need to specify data type, because Python automatically can convert according to the assigned value.

Keywords can be listed using the given line of code, "exec'" is no longer a keyword:

```
>>>import keyword
>>>print(keyword.kwlist)
['False', 'None', 'True', 'and', 'as', 'assert', 'async', 'await', 'break',
'class', 'continue', 'def', 'del', 'elif', 'else', 'except', 'finally',
'for', 'from', 'global', 'if', 'import', 'in', 'is', 'lambda', 'nonlocal',
'not', 'or', 'pass', 'raise', 'return', 'try', 'while', 'with', 'yield']
>>>x=10
>>>id(x)
4518782192d
```

2.2 Data types

There are two types of data in Python – mutable data type and immutable data type. Mutable data type can make any changes during the execution of the program while immutable data types cannot have any changes; instead, they always allocated a new memory location.

As we know, Python is an autotype caste language, but in the background, it manages the various data types to store different types of values. Data type of any variable can be visualized by method type() as given in the snapshot:

```
>>>message = "Good Morning!"
>>>print("message data type is ", type(message))
message data type is <type 'str'>
>>>message= "Hi, There"
>>>message
Hi, There
>>>age = 20
>>>print( "age data type is ", type(age))
age data type is <type 'int'>
```

Table 2.2: Data types.

Property	Behavior	Type	Example
Immutable	Numeric	Integer Float	x = 25 pi = 3.14
		Complex	complex_num=3.14j
		Long	long_num = 343445435453L
	Sequence	String	fname = "Tarkeshwar" lname = "Barua" nameInHindi=u" "
		Byte	byteString = b"Tarkeshwar"
Mutable		Byte array	byteArrayString = bytearray(b"Tarkeshwar")
		List	testList = [1,2,3,4,5,5,2,5,5,2,3,4]
Immutable		Tuple	testTuple = ("Francis", "Patrick", "Nixon")
Mutable	Sets	Set	fullname = set("Tarkeshwar Barua")
Immutable		Frozen set	testFrozenSet = (["James", "Sam", "Dennis", "Toney"])
Mutable	Mapping	Dictionary	student = {"name":"Tarkeshwar Barua", "address": "123 Monrovia", "phone":123456789}

```
>>>height = 5.6
>>>print( "height data type is ", type(height))
height data type is <type 'float'>
>>>paid = True
>>>print( "paid data type is ", type(paid))
paid data type is <type 'bool'>
>>> x="Greeting"
>>> type(x)
<class 'str'>
>>> x=True
>>> type(x)
<class 'bool'>
>>> x=3.1343434
>>> type(x)
<class 'float'>
```

2.2.1 Booleans

Boolean data type stores value as either True or False. Logical operations like **and, or, not,** can be performed on Boolean. Boolean can be integer values too, which stand as 1 means True and 0 means False:

```
>>> x=True      # boolean assignment to variable
>>> y=False
>>> type(x)
<class 'bool'>
>>>a and b      #if x is True and y is True then result will be TRUE
>>>a or b       #if x is False and y is True then result will be TRUE
>>>a not b      #if x is True and y is False then result will be TRUE
```

2.2.2 Integer

Integer data type is used to store the fractional value of a number. In another way, it stores only value without a decimal point:

```
>>>x=10
>>>type(x)
<class 'int'>
```

2.2.3 Float

Float data type is used to store a decimal value with period. In another way, it stores only value with decimal point:

```
>>>x=10.4434343
>>>type(x)
<class 'float'>
```

2.2.4 String

String data type is used to store a string (a sequence of characters) or a paragraph:

```
>>>name = "Tarkeshwar Barua"        #variable declaration
>>>type(name)                # type of variable
<class 'str'>
```

```
>>>name[0]                 #slicing from string
'T'
>>>name[0:10]              #slicing from String
'Tarkeshwar'
```

2.2.5 None

None stands for void or null. This type of data type is used to clear value from the given variable or memory location:

```
>>>name = None
>>>type(name)
<class 'NoneType'>
```

2.3 Block indentation

Like other programming languages such as Java, C, and C++, it does not have block creation using curly bracket {}. In Python, tabs are being used to create block so that we can control and construct loop. Here, the programmer has used white space carefully, and a wrong calibration may cause an error. Python uses colon (:) and indentation for showing where blocks of code begin and end. If the line following a colon is not indented, wrong indentation will raise **IndentationError** as in the following example:

```
>>> if 3>4:
print("Hello")
SyntaxError: expected an indented block
```

We can write code in server way, some of them are illustrated in Table 2.3.

Table 2.3: Block indentation.

if 5 < 2:	**if** 5 < 2: print(**"Hello"**)
print(**"Hello"**)	**else**: print(**"World"**)
else:	
print(**"World"**)	

Here, both examples are correct but the second example is not considered as good practice; therefore, we should avoid this practice. We should always use four spaces for indentation. python3 does not allow mixing the use of tabs and spaces for indentation:

```
>>>def myFunction():
. . . a=10
. . . return a
>>>print(myFunction())
```

2.4 Math calculations

Maths operations are very important in any programming language. Python also has an individual math module to perform complex mathematical operations. This needs to be imported from standard library to our program. Python has a bunch of built-in arithmetic operators; Math module contains implementations of common mathematical operations such as square root:

```
>>>3*6+3+4*(6+3)
57
>>>3*6+2-3*(6+4)-5
-15
>>>2**3+8 - 2**2+16
28
>>>2**2*4 /16%15 + 1 -3
-1.0
>>>import math
>>>math.sin(math.radians(30))
4.99999999999994
>>>math.cos(math.radians(30))
0.8660254037844387
>>>math.tan(math.radians(30))
0.5773502691896256
>>>math.pi
3.141592653589793
>>>math.e
2.718281828459045
>>>math.sqrt(64)
8.0
```

2.5 Operators

Operators are special symbols that represent computations such as arithmetic, logical, assignment, increment, and decrement operator. The value that is applied to an operator is called operands. These calculations are done by BODMAS, which stands for the bracket, of, division, multiplication, addition, subtraction, where (**) means power 10^3:

```
>>>print(10+30*50-40/20)
1508.0
>>>print(10+30*(50-40)/20)
25.0
>>>print(10+30*(50-40)**3/20)
1510.0
```

Parentheses have highest precedence and can be used to force an expression to evaluate in the order we want, and subtraction has lowest precedence.

2.6 String operations

A string can be created by enclosing any word character or number within single and double quotes. A string is a group of numbers, characters, and special symbols. The process of adding two or more than two strings is called concatenation. A string can be created as follows:

```
>>>firstName= "Tarkeshwar"
>>>type(firstName)
<type 'str'>
>>>lastName= "Barua"
>>>type(lastName)
<type 'str'>
>>>fullName=firstName+ " " +lastName
Tarkeshwar Barua
>>>type(fullName)
<type 'str'>
----------------above given program using PyCharm IDE--------------------
string="tarkeshwar barua"
print("Camel Case - "+string.capitalize()) # camel case
string1="TARKESHWAR"
print("Lower Case - "+string1.lower()) #lower case
```

```
print("Upper Case - "+string.upper()) #upper case
print("Replace A with B - "+string.replace("a","b")) # replace String
splitedword=string.split(" ") #split a string with space
for s in splitedword: #iterating splited words
    print("Splited Word is - "+s)
print("Case fold - "+string1.casefold()) #Return a version of the string suit-
able for caseless comparisons
print("Swap case - "+string1.swapcase()) #Convert uppercase characters to
lowercase and lowercase characters to uppercase
string2="data data data"
print("Find A - ",string2.find("d")) #Return the lowest index in S where sub-
string sub is found, such that sub is contained within S[start:end]. Optional
arguments start and end are interpreted as in slice notation
```

2.7 Indexing

A string is a sequence of characters. We can access any character in a Python string using its index, either counting from the beginning or from the end of the string. This indexing begins from 0(zero), as given in Table 2.4:

Table 2.4: String indexing.

T	A	R	K	E	S	H	W	A	R		B	A	R	U	A
0	1	2	3	4	5	6	7	8	9	10	11	12	13	14	15
−16	−15	−14	−13	−12	−11	−10	−9	−8	−7	−6	−5	−4	−3	−2	−1

```
>>>name = "TARKESHWAR BARUA"
>>>name[0]
'A'
>>>name[11]
'B'
>>>name[-3]
'R'
>>>name[-13]
'K'
```

2.8 Slicing

A sequence of characters can be sliced from the given string by specifying a range of indexing number; this process is known as Slicing. This will be a shorter string from the long string. Let us have a look in the following example:

```
>>>name(name)
16
>>>name[0:10]
'TARKESHWAR'
>>>name[11:16]
'BARUA'
>>>name[-16:-6]
'TARKESHWAR'
>>>name[-5:-1]
'BARU'
```

2.9 Concatenation

The process to join two or more than two strings together is called concatenation. This can be performed by "+" plus sign between two strings as in the given example:

```
>>>greeting = "Good Morning"
>>>fname= "Tarkeshwar"
>>>lname= "Barua"
>>>print(greeting+fname+lname)
'Good MorningTarkeshwarBarua'
>>>fname*3
'TarkeshwarTarkeshwarTarkeshwar'
```

2.10 Single-line and multiline comments

Nobody is perfect, that is why we may require deactivating some code for trouble-shooting purposes, and this commenting can make the code more human-understandable. As the code size gets bigger, it becomes more complicated for reading, code understanding, and debugging. Then comments play a vital role to explain the code functionality in natural language. These comments can be single line or multiline. Single-line comment can be given by using # while multiline comment needs triple quote. Comments are very useful while documenting nonobvious features of the code. Let us have a look at the following example:

```
>>>def myFunction():              #this is single-line comment
. . . a=10
. . . return a
```

"this is a multiline comment example which can be achieved by using triple quote open and close and this is not a part of code, it is only for information to programmer":

```
>>>print(myFunction())
```

2.11 Functions

A function is a named sequence of statements that is designed to perform a certain task, or function is a method that is stored as a class attribute. Function returns a value based on its given arguments. We can create it by using "**def**" keyword and function name then we can write the number of lines to execute in the function. Later, this function can be called by its name to execute its written statement. There are some rules to define a function as follows:

1) A function cannot start with any special symbol(`, ~, !, @, #, $, %, ^, &, *, -, +, =, ", ", <, >, /, ?,).
2) A function should not have a special symbol (., @).
3) A function can not start with a number.
4) A function can start with _ (underscore) symbol.
5) A keyword cannot be a function name.

A function definition specifies the name of a new function and the sequence of statements that execute, when the function is called. There are two types of functions as follows:

2.11.1 Built-in functions

Built-in functions are already created in python and there is no need to create them. Built-in functions can be listed using **dir(__builtins__)** function. The functionality of any function can be listed by **help(max)** function. We can follow the examples where id is a built-in function of Python:

```
>>>x=10
>>>id(x)
4504909200
>>>pow(2,3)
>>>dir(__builtins__)
'ArithmeticError',   'AssertionError',   'AttributeError',   'BaseException',
'BlockingIOError',   'BrokenPipeError',   'BufferError',   'BytesWarning',
'ChildProcessError',        'ConnectionAbortedError',        'ConnectionError',
'ConnectionRefusedError',   'ConnectionResetError',   'DeprecationWarning',
'EOFError',   'Ellipsis',   'EnvironmentError',   'Exception',   'False',
'FileExistsError',        'FileNotFoundError',        'FloatingPointError',
'FutureWarning', 'GeneratorExit', 'IOError', 'ImportError', 'ImportWarning',
'IndentationError', 'IndexError', 'InterruptedError', 'IsADirectoryError',
'KeyError',        'KeyboardInterrupt',        'LookupError',        'MemoryError',
'ModuleNotFoundError',     'NameError',     'None',     'NotADirectoryError',
'NotImplemented',   'NotImplementedError',   'OSError',   'OverflowError',
'PendingDeprecationWarning',   'PermissionError',   'ProcessLookupError',
'RecursionError',   'ReferenceError',   'ResourceWarning',   'RuntimeError',
'RuntimeWarning',  'StopAsyncIteration',  'StopIteration',  'SyntaxError',
'SyntaxWarning',  'SystemError',  'SystemExit',  'TabError',  'TimeoutError',
'True',   'TypeError',   'UnboundLocalError',   'UnicodeDecodeError',
'UnicodeEncodeError',        'UnicodeError',        'UnicodeTranslateError',
'UnicodeWarning', 'UserWarning', 'ValueError', 'Warning', 'WindowsError',
'ZeroDivisionError', '_', '__build_class__', '__debug__', '__doc__', '__im-
port__', '__loader__', '__name__', '__package__', '__spec__', 'abs', 'all',
'any', 'ascii', 'bin', 'bool', 'breakpoint', 'bytearray', 'bytes', 'callable',
'chr', 'classmethod', 'compile', 'complex', 'copyright', 'credits', 'delattr',
'dict', 'dir', 'divmod', 'enumerate', 'eval', 'exec', 'exit', 'filter',
'float', 'format', 'frozenset', 'getattr', 'globals', 'hasattr', 'hash',
'help', 'hex', 'id', 'input', 'int', 'isinstance', 'issubclass', 'iter', 'len',
'license', 'list', 'locals', 'map', 'max', 'memoryview', 'min', 'next', 'ob-
ject', 'oct', 'open', 'ord', 'pow', 'print', 'property', 'quit', 'range',
'repr', 'reversed', 'round', 'set', 'setattr', 'slice', 'sorted', 'staticme-
thod', 'str', 'sum', 'super', 'tuple', 'type', 'vars', 'zip']
```

2.11.2 User-defined functions

User-defined functions are created by the programmer to perform some task. Let us follow the examples:

```
>>>def myFunction():          #defining a customized function
. . . a=10
```

```
. . . return a
>>>print(myFunction())        #calling defined function
>>>def greeting():
. . . print("Good Morning Friends")      #printing output inside function
>>>print(greeting())
```

2.11.3 Function with argument

Function can be created with an argument too. It is very useful to pass value inside the function to achieve dependency injection. After this injection, function works independently without its outside interaction. It may return some output or may not, which depends upon the system requirement. In the example, we have passed two values x and y as function parameters; after passing, these values can perform the task independently as a local variable for this function and they can produce output. In the second line, we have printed output and the second-line method is returning value out of function. Let us study the given example:

```
>>>def addition(x, y):
. . . print("passed value result is ", (x+y)) #printing output inside function
. . . return x + y                     #returning value out of function
>>>print(addition(10, 20))
```

2.11.4 Flow of execution

Function should be defined before its execution because Python is an interpreted programming language. It interprets function definition and then the execution is possible. This sequence is called the flow of execution. Execution begins with the first statement of the program. The statement inside the function is not executed until the function is called.

2.11.5 Default values

Sometimes it happens that a function needs to pass a value as a parameter. Unfortunately, if we forget to pass those values it may cause the error in the program and, as a result, the program stops executing. To avoid this situation, python provides a default value feature; in case the programmer doesn't provide any value, it will take the default value. Passing default value also has certain rules. The second parameter can not be dynamic while first is the default. The first parameter should be dynamic but the other can be the default if a function has more than one parameter as we can see in the example:

```
>>>def greeting(name= "Guest"):
   print("Hello Mr. ",name)
>>>greeting()
Hello Mr. Guest
>>>greeting("Tarkeshwar Barua")
Hello Mr. Tarkeshwar Barua
>>>def greeting1(fname= "Tarkeshwar", lname= "Barua"):
   print("Hello Mr. ",fname, lname)
>>>greeting1()
Hello Mr. TarkeshwarBarua
>>>def greeting2(fname= "Tarkeshwar", lname):
   print("Hello Mr. ",fname, lname)
  File "<stdin>", line 1
SysntaxError: non-default argument follows default argument
```

2.12 Return value

Python is a dynamically typed casting language because of which no return type is needed to be specified during the function definition. Following is an example if we want to encapsulate this logic in a function:

```
>>>def addition(a, b):
   return a+b
>>>x=addition(3,6)
>>>type(x)
<class 'int'>
>>>x
 9
>>>y=addition("Tarkeshwar", "Barua")
>>>type(y)
<class 'str'>
>>>y
'TarkeshwarBarua'
```

We have seen that the same function can work on string as well as numbers because it is autotype cased.

2.13 Type-casting

Python provides built-in functions that convert values from one type to another type. Let us assume we want to convert from **string** to **int,** following the given example:

```
>>>int('15')
15
>>>int('Hello World')
ValueError: invalid literal for int() : Hello World
>>>int(3.1456453)
3
>>>str(3.1456453)
'3.1456453'
```

2.14 Disassembler

We can write disassembler module in short as **dis.** Byte code is an implementation of a Cython interpreter. It is used for analysis of Cython byte code by dissembling. The byte code analysis API allows a piece of Python code to be wrapped into a byte code object. Cython byte code is an input for this module in the included opcode.h. Let us see the example:

```
import dis
def greeting(name="Guest"):
    return "Hello, Mr "+name+"!"
print(greeting())
print(greeting("Tarkeshwar"))
dis.dis(greeting)
-------------------------------output-------------------------------
Hello, Mr Guest!
Hello, Mr Tarkeshwar!
5       0 LOAD_CONST      1 ('Hello, Mr ')
        2 LOAD_FAST       0 (name)
        4 BINARY_ADD
        6 LOAD_CONST      2 ('!')
        8 BINARY_ADD
        10 RETURN_VALUE
-------------------------------------------------------------------
```

2.15 Lambda keyword

Lambda keyword provides a shortcut for declaring small and anonymous functions. Lambda functions are single-expression functions that are not necessarily bound to the name, which means they are anonymous. Lambdas cannot use regular python statements and always include an implicit **"*return*"** statement. These functions can have any number of arguments but only one expression that is evaluated and returned. It is syntactically restricted to a single expression as follows:

```
def add(x,y):          #an ordinary function
    return x+y
sub=lambda x,y:x-y        #A Lambda Function
print(add(2,2))        #lambda function calling
print(sub(4,2))        #Ordinary function calling
print((lambda x,y:x*y)(3,4)) #lambda with *function expression
```

Filter function takes in a function and a list of argument and can return group of item after performing certain task over it. Let us see in the code:

```
multiple=list(filter(lambda x:(x>10),items)) #being copied in to list, those
items are greater then 10
print(multiple)
print(type(multiple))
```

The above-given code can be used with map function too as it is given in the next code:

```
multiple=list(map(lambda x:(x>10),items)) #being copied in to list, those
items are greater then 10
print(multiple)
print(type(multiple))
```

Lambda function can return any type of function, let us see in the code:

```
from functools import reduce
items =[34,3,42,4,23,4,2,3,43]
multiple = reduce((lambda x, y: x * y), items)
print (sum)
```

Summary

- Python is a convenient and easy programming language to understand and develop an application in a short period. Python comes with basics, all data types, variables, and so on. Python is widely used among programmers.
- Python is free to use with big community support. It is a very good programming language in terms of memory management by garbage collectors.
- Python basics include data type variables, and its type conversions and various operators. Some unique words that have some special meaning for Python are called keywords.
- Python does not have curly bracket { } to create blocks. It needs to be managed by indentation. We can perform mathematical calculations efficiently with the math module – various types of operators such as arithmetical, logical, comparison, and assignment.
- The string is an immutable object in all the programming languages and the same is applied for Python too. It means that a program cannot be modified at run time of program still if we want to perform some task; it is called concatenation and slicing to get sub-string from the big string.
- Comments can be made in two ways, single-line comment by applying # before the line and multiline can be done by enclosing with three single/double quotes.
- Functions are a combination of statements that execute while they have been called. It makes programming very easy to maintain. One function can be called several times; in python, functions support default value too.
- Conversion from one data type to another, the process is known as data type casting. Python supports autotype casting, but it is also possible explicitly.

Key terms

Python is an autotype casting language, that is why, it is not needed to specify a variable type for variable, according to the stored value its type changes. The keyword cannot be used as a variable name. Blocks are maintained by indentation. Mathematical calculations can be performed very quickly with built-in library math. Operators are used to perform operations over operands. Indexing always begins with zero, and it goes up to length −1. On behalf of this string, operations like slicing and concatenation are performed. Comments are used to deactivate and provide some information related to programming code.

Review questions

1. Execute the given code and find out the output
   ```
   width = 30
   height = 20
   length = 50
   area = width*height*length
   perimeter = width + height + length
   ```

2. Find out the area and perimeter of a circle while its radius is 3.5 Mtr
   ```
   radius = 3.5
   pi=3.14
   area = pi*r*r
   perimeter = 2*pi*r
   ```

3. Let us assume that the base price of 1 kg apple is 100 INR, but fruit seller got 10% discount while he/she purchased and 3% freight charge to carry in shop. What is the sale price if he wants to make 10% profit on it?
   ```
   >>>baseprice = 100
   >>>puchase_discount = 100*(10/100)
   >>>purchase_price = baseprice - purchase_discount
   >>>freight= purchase_price*(3/100)
   >>>total_purchase_price = purchase_price + freight
   >>>sale_price = total_purchase_price + total_purchase_price * (10/100)
   ```

4. Explain different data types of Python.

5. What type of functions are available in Python? Explain.

Exercise

Tick the correct option

Q.1. Is Python case sensitive while dealing with identifiers?
   a) Yes
   b) No
   c) Machine dependent
   d) None of the above

Q.2. What is the maximum possible length of an identifier?
   a) 31 characters
   b) 63 characters
   c) 79 characters
   d) None of the above

Q.3. Which of the following is an invalid variable?
   a) my_string_1
   b) 1st_string
   c) foo
   d) __

Q.4. Which of the following is not a keyword?
   a) eval
   b) assert
   c) nonlocal
   d) pass

Q.5. All keywords in Python are _____
   a) lowercase
   b) upper case
   c) capitalized
   d) none of the above

Q.6. Which of the following is true for variable names in Python?
   a) Unlimited length
   b) All private members must have leading and trailing underscores
   c) Underscore and ampersand are the only two special characters allowed
   d) None of the above

Q.7. Which of the following is an invalid statement?
   a) abc=1,000,000
   b) a b c=1000 2000 3000
   c) a, b, c=1000, 2000, 3000
   d) a_b_c=1,000,000

Q.8. Which of the following cannot be a variable?
   a) __init__
   b) in
   c) it
   d) on

Q.9.   print("Hello {0!r} and {0!s}".format("foo", "bin")):
   a) Hello foo and foo
   b) Hello "foo" and foo
   c) Hello foo and "bin"
   d) Error

Q.10.   Which of the following data types is not supported in Python?
   a) Numbers
   b) String
   c) List
   d) Slice

Answers

Q.1.   b) Yes
Q.2.   d) None of the above
Q.3.   b) 1st_string
Q.4.   a) eval
Q.5.   d) None of the above
Q.6.   a) Unlimited length
Q.7.   b) a b c=1000 2000 3000
Q.8.   b) in
Q.9.   b) Hello "foo" and foo
Q.10. d) Slice

Fill in the blanks

1.   _____are some special characters that have some special meaning for programming language.
2.   _____data type can make any changes during the execution of the program.
3.   _____is the process to join two or more than two strings together.
4.   _____is a sequence of statements that is designed to perform a certain task.
5.   Sequence of character from the given String is known as _____.

Answers

1. Keywords
2. Mutable
3. Concatenation
4. Function
5. Slicing

3 Conditions and loops

> Most good programmers do programming not because they expect to get paid or get adulation by the public, but because it is fun to program. ~Linus Torvalds

3.1 If and else and conditional statements

Sometimes we need to make a decision like "if" and "else" in the programming. There are a lot of uses for the simple statements and operations we have covered so far in this book. Python provides if–else, and elif keywords to do so. However, the main purpose of any programming is to implement logic. If/else blocks are core syntax elements in Python. They let us conditionally control the flow of the program. For instance, once our car is running very fast and we have exceeded the limit 100 km/h, then it should print "You are Driving too Fast"; otherwise, it should print "You are driving under the limit, Well done" to get us be aware, it goes as:

```
>>>def carDriving(carSpeed):
. . . if carSpeed > 100:      #checking condition true/false
. . .         print("You are Driving too fast")
. . . else:                   #execute statement if condition is false
. . .         print("You are driving under limit, Well done")
>>>carDriving(105)
You are driving too fast
>>>carDriving(90)
You are driving under limit, Well done
```

In Python, any value can be interpreted as True or False. Among the data types value 0, empty string, none, and False. Other values are considered as True. We can check if the value is True or False by converting the value to a Boolean type as given in the code:

```
>>>4<5
True
>>>5<4
False
>>>3<=3
True
>>>4<=3
False
>>>2>1
True
```

https://doi.org/10.1515/9783110689488-003

```
>>>not None
True
>>>3>3
False
>>>2>=2
True
>>>3>=4
False
>>> "Tarkesh"== "Tarkesh"
True
>>>0==1
False
>>>not True
False
>>> "Tarkesh" != "Barua"
True
>>>10 !=(9+1)
False
>>>3 and "Barua"
'Barua'
>>>1 or None
1
>>>None or False
False
```

Comparison operators have higher precedence than logical operators like and/or/ not. Math operators have higher precedence than comparison operator, as we can see in the example:

```
>>>3*6+3<4*(6+3)
True
>>>3*6+2==3*(6+4)-5
False
>>>2**3==8 and 2**2==16
False
>>>(2**2==4 and False) or (not(16<15 or 1 ==3))
True
>>>def result(marks):
. . . if marks >= 75:
. . .     return "Distension"
. . . elif marks >= 60:
. . .     return "First Division"
```

```
. . . elif marks >= 45 and marks <60:
. . .      return "Second Division"
. . . elif marks >= 33:
. . .      return "Third Division"
. . . else:
. . .      return "Fail"
>>>result(77)
Distension
>>>result(74)
First Division
>>>result(59)
Second Division
>>>result(44)
Third Division
>>>result(32)
Fail
```

3.1.1 Nested if–else

Sometimes logic can be more sophisticated than just two conditions. Then we need to apply nested if-else, as we can see in the code, we have passed two conditions in the same condition (inline condition) using and keyword. Let us see in the given code:

```
def identify_num(num):
    if num >= 0:
        if num == 0:
            print("Zero")
        else:
            print("Positive number")
    else:
        print("Negative number")
input_num = float(input("Enter a number: "))
identify_num(input_num)
```

```
---------------------------------Output---------------------------------
Enter a number: -3
Negative number
-----------------------------------------------------------------------
Enter a number: 0
Zero
```

```
-------------------------------------------------------------------------
Enter a number: 3
Positive number
-------------------------------------------------------------------------
```

3.2 Loops

In python, there are two types of loops "while" and "for." While carries on repeating as long as a certain condition is true. Statement inside a while loop must be intended, the loop must be defined by while keyword and Boolean statement and colon (:). Boolean statement can be a function call, variable, and comparison statement too, while provides break and continue statements to have even better control on our loop:

```
>>>num=0
>>>while num < 100:
    num += 1
    print(num)                      #prints numbers 0 to 100 as syntax is given
while <condition1>:
    <statement1>
    if <condition2>:
        <statement2>
        break
    if <condition3>:
        <statement3>
        continue
    else:
        <ststement4>
```

3.2.1 For loop

We can use **"for"** loop to iterate through a collection, as we can see in the example:

```
>>>friends=["James", "Sam", "Estherlyn", "Elizabeth"]
>>>for friend in friends:
. . . print(friend)
. . .
James
Sam
Estherlyn
```

```
Elizabeth
>>>student={'name': 'Tarkeshwar', 'age':32, 'height': 5.6, 'isGraduate':
True}
>>>for info in student:
. . . print(info, " - ", student[info])
. . .
name - Tarkeshwar
age - 32
height - 5.6
```

3.2.2 While loop

We can use the **"while"** loop to repeat specific action up to a specified number of times. We can control while loop by providing condition along with while keyword. We can increase or decrease counter using x = x + 1, x = x-1, where x is the counter variable. Let us see an example:

```
<initialization of variable>
while(condition):
    <statements>
    increment or decrement operator

>>>num=0
>>>while(num<100):
. . . print(num)
. . . num=num+1
. . .6
```

3.3 Break and continue statement

We can escape from a loop using break command and the loop can be continued by continue command. Given code will not proceed after the third item because it contains 8, which is a larger number than 6; hence, it will stop looping:

```
>>>numbers=[1,2,5,8,3,5,8,3,7,3,7,8,4,3]
>>>for number in numbers:
    print("Looking at: "+str(number))
    if number > 6:
        print("Too big: "+str(number) + "!")
        break
```

This iteration can be continued even after the break, continue statement:

```
>>>numbers=[1,2,5,8,3,5,8,3,7,3,7,8,4,3]
>>>for number in numbers:
. . . print("Looking at: "+str(number))
. . . if number > 6:
. . .      print("Too big: "+str(number) + "!")
. . .      break
. . .
```
Looking at: 1
Looking at: 2
Looking at: 5
Looking at: 8
Too big: 8!

This iteration can be continued even after break, continue statement:

```
>>>numbers=[1,2,5,8,3,5,8,3,7,3,7,8,4,3]
>>>for number in numbers:
. . . if number > 6:
. . .      continue
. . . print("Looking at: "+str(number))
. . .
```
Looking at: 1
Looking at: 2
Looking at: 5
Looking at: 3
Looking at: 5
Looking at: 3
Looking at: 3
Looking at: 4
Looking at: 3

3.4 Collections

Whenever it comes to store objects, their collection comes in the light, which provides a flexible way to store and it allows to store various operations over it using various built-in functions. For example, we might want to iron our socks. It would not make sense for us to consider each one individually as a unique object requiring special treatment. It would be very handy if we take them as a collection. Collection is a set of lockers. Python provides two types of collection: one with a name tag and the second one with anonymous. These are indexed by the numbers. If we wish to access the item, then the indexing number is a good approach. This type of collection is known as list and tuple. Collection with name, stores objects with name, these are known as dictionaries. It works on FIFO(First In First Out) and LIFO(Last In First Out). Collection can make our task easy, let us see in the given example of code, here we can find the frequency of the characters in the given character with few lines of code:

```
import collections
collection=collections.Counter("Tarkeshwar Barua")
print(collection)

Output-
Counter({'a': 4, 'r': 3, 'T': 1, 'k': 1, 'e': 1, 's': 1, 'h': 1, 'w': 1, ' ': 1,
'B': 1, 'u': 1})
```

3.4.1 Tuples

Tuples are immutable arrays of variable. Elements from the tuple can be accessed using index number and tuples are immutable, means no modification is possible after creation. We can perform slicing, indexing, concatenation, and multiply on tuples. To create tuples, objects must be surrounded by parentheses and separated by commas. Even tuple that has a single item must have comma ("Tarkeshwar",) as in the given code:

```
>>>friends=("Sam", "William", "James", "Lionel", "John")
>>>type(friends)
<class 'tuple'>
>>>friends
('Sam', 'William', 'James', 'Lionel', 'John')
>>>friends[0]
'Sam'
>>>friends[4]
'John'
>>>friends[1][0:4]
'Will'
>>>friends[1:3]
('William', 'James')
>>>friends[2]= "Elizabeth"          #tuple is immutable, no modification
File "<stdin>", line 1, in <module>
TypeError: 'tuple' object does not support item assignment
```

```
>>>friends[0][0:5]= "Elton"          #String is immutable, no modification
File "<stdin>", line 1, in <module>
TypeError: 'str' object does not support item assignment
>>>len(friends)
5
>>>friends*3     #creating new tuple with three time copy of friends tuple
('Sam', 'William', 'James', 'Lionel', 'John', 'Sam', 'William', 'James',
'Lionel', 'John', 'Sam', 'William', 'James', 'Lionel', 'John')
>>>friends+("Thomas","Reuben","David") #tuple concatenation of tuple
('Sam', 'William', 'James', 'Lionel', 'John', 'Thomas', 'Reuben', 'David')
```

Table 3.1: Functions of tuple data type.

Method name	Use	Explanation
Index	mytuple.index(item)	Return the first index number of the item.
Count	mytuple.count(item)	Return the number of the occurrences of the item.

3.4.2 Lists

Lists are mutable arrays of variable. Indices for lists (sequences) start counting with 0. Elements from the list can be accessible using index number; because lists are mutable – modification is possible after creation or at any moment of execution. We can perform slicing, indexing, appending, and multiply on lists. The slice operation, mixedList[1:3], return a list of items starting with indexed by one up to not including the item indexed 3 to 1. To create lists, objects must be surrounded by square brackets and separated by commas. Even lists that have a single item must have comma["Tarkeshwar",]. We can follow the given code:

```
>>>(54).__add__(21)
75
>>>[123, 'abcd', 3.14, 'd']
[123, 'abcd', 3.14, 'd']
>>>list(range(10))
[1, 2, 3, 4, 5, 6, 7, 8, 9]
>>>list(range(5,10)
[5, 6, 7, 8, 9]
>>>list(range(5,10,2)
[5, 7, 9]
>>>list(range(10,1,-1)
[10, 9, 8, 7, 5, 4, 3, 2]
```

```
>>>mixedList=[123, 'abcd', 3.14, 'd']        # it can store mixed data too
>>>friends=["Sam", "William", "James", "Lionel", "John"]
>>>type(friends)
<class 'list'>
>>>friends
['Sam', 'William', 'James', 'Lionel', 'John']
>>>friends[0]
'Sam'
>>>friends[4]
'John'
>>>friends[1][0:4]
'Will'
>>>friends[1:3]
('William', 'James')
>>>friends[2]= "Luke"              #list is mutable, third item will replaced
>>>friends
['Sam', 'William', 'Luke', 'Lionel', 'John']
>>>friends[0][0:5]= "Prince"        #String is immutable, no modification
File "<stdin>", line 1, in <module>
TypeError: 'str' object does not support item assignment
>>>len(friends)
5
>>>friends*3        #creating new list with three time copy of friends list
['Sam', 'William', 'James', 'Lionel', 'John', 'Sam', 'William', 'James',
'Lionel', 'John', 'Sam', 'William', 'James', 'Lionel', 'John']
>>>friends+["Thomas","Reuben","David"] #list concatenation of list
['Sam', 'William', 'James', 'Lionel', 'John', 'Thomas', 'Reuben', 'David']
>>>friends.pop()       #last item(David) will be deleted
'John'
>>>friends.pop(2)        #second idem(James) will be deleted
'James'
>>>friends.append("Emmanuel")        #item will be added at after last index
['Sam', 'William', 'Lionel', 'Emmanuel']
>>>friends.insert(2, "Taddy")        #Taddy will be added at 2 index and other
shifted to right side
['Sam', 'William', 'Taddy', 'Lionel', 'Emmanuel']
```

Table 3.2: Functions of list data type.

Method name	Use	Explanation
append	MixedList.append(item)	Adds a new item to the end of a list.
insert	MixedList.insert(i, item)	Adds a new item to the ith position in the list.
pop	MixedList.pop()	Removes and return last item of the list.
pop	MixedList.pop(i)	Removes and return i[th] item of the list.
sort	MixedList.sort()	Modifies a list to be sort.
reverse	MixedList.reverse()	Modifies a list to be sorted in reverse order.
del	del MixedList[i]	Removes and return i[th] item of the list.
index	MixedList.index(item)	Return the first index number of the item.
count	MixedList.count(item)	Return the number of the occurrences of the item.
remove	MixedList.remove(item)	Removes the first occurrence of the item.

3.4.3 Sets

Sets are unordered collections and mutable array of variables. Duplicate elements are removed automatically. Sets are written as comma-delimited values enclosed in curly braces. They are used in cases where they need to be grouped together, and not in what order they are included. The empty set is represented by set(). For a large group of data, it is much faster to check whether or not an element is in a set than it is to do the same for the list. Sets are heterogeneous, and the collection can be assigned to a variable:

```
>>>a=set("Tarkeshwar Barua")      #all duplicates are removed
>>>a
{'h', 'u', 's', 'w', 'r', 'a', 'B', 'e', ' ', 'k', 'T'}
>>>names={"Sam", "William", "James", "Lionel"}
>>>names
{'Sam', 'William', 'James', 'Lionel'}
>>>len(names)
4
>>>type(names)
<class, 'set'>
>>>mylist=[1,2,3,4,5,6,7,7,4,2,2]
>>>myset=set(mylist)
>>> data={3,6,"cat",4.5,False}
>>>False in data
True
>>>"cat" in data
True
```

Table 3.3: Functions of set data type.

Method name	Use	Explanation
union	set1.union(set2)	Returns a new set with all elements from both sets.
intersection	set1.intersection (set2)	Returns a new set with only the elements common to both sets.
difference	set1.difference (set2)	Returns a new set with all elements from first sets and those are not found in second.
issubset	set1.issubset (set2)	Asks whether all elements of one set are in the other.
add	set1.add(set2)	Adds items to the set.
pop	set1.pop()	Removes item from set.
clear	set1.clear()	Removes all the elements of the set.
remove	set1.remove (item)	Removes the occurrence of the item.

Sets do not support the sequential mechanism but support few familiar operations such as union, intersection, issubset, and difference.

3.4.4 Frozen sets

Frozenset() function returns an immutable set. It is an immutable version of the set object. It does not provide auto-indexing of its elements like a set, which means any element can dynamically occupy memory space anywhere in the memory that causes different outputs at different times. Frozen sets are unordered collections and immutable array of variables. Duplicate elements are removed automatically. As we have studied in the set, we can modify at any moment but this is not possible in the frozen set after creation. Frozen set can be used as a key or a value in dictionaries. Let us see in the given example:

```
>>>a=frozenset(["New Delhi", "Mumbai", "Kolkata", "Chennai"])
>>>a
frozenset({'Kolkata', 'New Delhi', 'Chennai', 'Mumbai'})
```

Table 3.4: Functions of frozen set data type.

Method name	Use	Explanation
union	set1.union(set2)	Returns a new set with all elements from both sets.
intersection	set1.intersection (set2)	Returns a new set with only the elements common to both sets.
difference	set1.difference (set2)	Returns a new set with all elements from first sets and those are not found in second.
issubset	set1.issubset (set2)	Asks whether all elements of one set are in the other.

3.4.5 Byte

Bytes are unordered collections. They contain a single immutable byte. Bytes can be created by using either byte constructor or b character just before the given string. Bytes are immutable array of variables. It prints as ASCII characters during print on screen. It has indexing and slicing behavior:

```
>>>a=b'Tarkeshwar Barua'
>>>a
b'Tarkeshwar Barua'
>>>c=bytes('Good Morning')
>>>c
b'Good Morning'
```

3.4.6 Byte array

Byte array is an array of byte data type. It has mutable property in a sequence of integers from 0 to 255. Byte arrays are unordered collection and mutable array of variables. The optional source parameter can be used to initialize the array in a few different ways:
1) In the case of a string, we have to provide its encoding parameters with function str.encode().
2) In the case of iterate, it must be within 0–255 integers.
3) If it's an integer, then the array will have that size and will be initialized with null byte.
4) If it's an object, then the read-only buffer of the object will be used to initialize the byte array.

Let us see the example:

```
>>>a=bytearray(b "Tarkeshwar Barua")
>>>a
bytearray(b'Tarkeshwar Barua')
```

3.4.7 String

Python does not have character data type that is why single character is also considered as a string. Strings can be created by either single or double and triple quotes too. We can create double-quote string inside the triple quote and a single quote string can be created inside double quotes. A triple quote is used to create a multiline string. Strings are immutable. It contains a Unicode character. Any individual character of the string can be accessed by using the function of indexing with range or without range. The range operator colon (:) symbol is used for slicing in string. String in a single quote cannot hold any other single-quoted character in it; that is why double quotes are preferred:

```
>>>fname= 'Tarkeshwar'        #creating string constant by single quote
>>>type(fname)
<class 'str'>
>>>print(fname)
Tarkeshwar
>>>lname= "Barua"        #creating string constant by double quote
>>>type(lname)
<class 'str'>
>>>print(lname)
Barua
>>>fname+ " "+lname
'Tarkeshwar Barua'
>>>address= '''House No 123,
abc colony,
monrovia'''        #creating string constant by triple single quote
>>>type(address)
<class 'str'>
>>>print(address)
House No 123,
abc colony
monrovia
>>>phone= """"+231-
. . .770264434"""        #creating string constant by triple double quote
```

```
>>>type(phone)
<class 'str'>
>>>print(phone)
+231-
770264434
>>> print("Hello","World", sep="***")
Hello***World
>>> print("Hello","World", end="***")
Hello World***
>>> print("Hello",end="***"); print("World")
Hello***World
```

Table 3.5: Functions of string data type.

Method name	Use	Explanation
center	a_string.center(w)	Returns a string centered in a field of size w.
ljust	a_string.ljust(w)	Returns a string left-justified in a field of w.
lower	a_string.lower()	Returns a string in all lower case.
rjust	a_string.rjust(w)	Returns a string right-justified in a field of w.
find	a_string.find(item)	Return the first index number of the item.
count	a_string.count(item)	Return the number of the occurrences of the item.
split	a_string.split(s_char)	Splits a string into sub-strings at s_char.

3.4.8 Dictionaries

Dictionaries are mutable arrays of variables. Elements are being stored in the form of key-value pair. Elements from the dictionary can be accessible using keys. As dictionaries are mutable; modification is possible after creation or at any moment of execution. We can perform slicing, appending, and multiply on dictionary. To create dictionaries, objects must be surrounded by curly bracket {} and separated by commas, and keys and values are separated by colon(:). Even dictionaries having single item must have comma ['name':"Tarkeshwar",]. Keys in dictionary should be unique but value can be same. A key can be any type of variable. During changes if a key does not exist then a new key will be created. It is an unordered collection of map implementation. We can follow the given code:

```
>>>student={'name': 'Tarkeshwar', 'age':25, 'height': 5.6, 'isGraduate': True}
>>>type(student.values())            # it will print all values
>>>type(student.keyss())             # it will print all keys
>>>type(student)
<class 'dict'>
```

```
>>>student
{'name': 'Tarkeshwar', 'age':25, 'height': 5.6, 'isGraduate': True}
>>>student['name']
'Tarkeshwar'
>>>student['isGraduate']                      #value can retrieved by key
True
>>>class Car(object):
    def __init__(self, model, color):
        self.model=model
        self.color=color
>>>mercedes=Car("E400", "black")                # Key could be any object
>>>testDict={mercedes:Car("E300", "Gray"), "BMW":Car("X6","Black")}
>>>testDict[mercedes]
<__main__.Car object at 0x10d038d30>
>>>testDict[mercedes].model
'E300'
>>>testDict["BMW"].color
'Black'
>>>testDict["BMW"]= "Emmanuel"   #dictionary is mutable, BMW key will replaced
>>>len(testDict)
2
>>>testDict*3      #multiplying dictionary not allowed
Traceback (most recent call last):
    File "<stdin>", line 1, in <module>
TypeError: unsupported operand type(s) for *: 'dict' and 'int'
>>>testDict.pop(mercedes)      #key can be removed
>>>testDict+{"name":"Parker", "age", "27")     #dictionary concatenation not
allowed
Traceback (most recent call last):
    File "<stdin>", line 1, in <module>
TypeError: unsupported operand type(s) for +: 'dict' and 'dict'
>>>testDict["height"]=5.6                  #inserting height in the dictionary
```

Table 3.6: Functions of dictionary data type.

Method name	Use	Explanation
keys	my_dict.keys()	Returns the keys of the dictionary in a dict_keys object.
values	my_dict.valiues()	Returns the values of the dictionary in a dict_values object.
items	my_dict.items()	Returns the key value pair in a dict_items object.
get	my_dict.get(key)	Returns the value associated to the key, otherwise none.
get	my_dict.get(key,alt)	Returns the value associated to the key, otherwise alt.

3.4.9 Arrays

In arrays, individual elements can be accessed through indexes. Python arrays are zero-indexed. An array is a data structure that stores values of the same type. This is the main difference between arrays and lists. We need to import the array module because the array is not a fundamental data type like others. Type codes are used to define the type of array values or the type of array:

```
>>>from array import *
>>>firstArray=array('i', [1,2,3,4,5,6,7])     #i (signed integer) is type code
[initializing values]
>>>firstArray
array('i', [1, 2, 3, 4, 5, 6, 7])
>>>type(firstArray)
<class, 'array.array'>
>>>firstArray[1]
2
>>>firstArray.append(8)     # appending new value in the array after last index
>>>firstArray.insert(3, 9)       #inserting values at third index
>>>firstArray
array('i', [1, 2, 3, 9, 4, 5, 6, 7, 8])
>>>secondArray=array('i', [10,12,13,14,15,16,17])
>>>firstArray.extend(secondArray)       # extending array using extend function
>>>firstArray
array('i', [1, 2, 3, 9, 4, 5, 6, 7, 8, 10, 12, 13, 14, 15, 16, 17])
>>>c=[18,19,20]
>>>firstArray.fromlist(c)       #adding list into array
>>>firstArray
array('i', [1, 2, 3, 9, 4, 5, 6, 7, 8, 10, 12, 13, 14, 15, 16, 17, 18, 19, 20])
>>>firstArray.remove(4)       #removing 4 index item
>>>firstArray
array('i', [1, 2, 3, 9, 5, 6, 7, 8, 10, 12, 13, 14, 15, 16, 17, 18, 19, 20])
>>>firstArray.pop()       #removing last index item
5
>>>firstArray
array('i', [1, 2, 3, 9, 6, 7, 8, 10, 12, 13, 14, 15, 16, 17, 18, 19, 20])
>>>firstArray.index(5)       #retrieving 5 index item
>>>firstArray.reverse()       #reverse entire array positions
>>>firstArray
array('i', [20, 19, 18, 17, 16, 15, 14, 13, 12, 10, 8, 7, 6, 9, 3, 2, 1])
>>>firstArray.count()       #returns length of the array
```

3.5 Date and time

Date and time associated with datetime module. datetime itself is one object that can be used to store date and time. It has these main factors:
- **date** – It includes day, month, and year.
- **time** – it includes hour, minutes, seconds, and microsecond.
- **datetime** – it includes day, month, year, hour, minutes, seconds, and microsecond.
- **timedelta** – a value that is used to manipulate dates.
- **tzinfo** – it is an abstract class which is responsible to handle time zone.

Python 3.2 and above has support for %z (timezone)format when parsing a string into a **datetime** object, and other libraries too are available such as dateutil:

```
>>>import datetime
>>>currentDateTime=datetime.datetime.strptime("2019-06-03T15:27:18-0530",
"%Y-%m-%dT%H:%M:%S%z")
>>>currentDateTime
datetime.datetime(2019, 6, 3, 15, 27, 18, tzinfo=datetime.timezone(datetime.
timedelta(days=-1, seconds=66600)))
>>>datetime.datetime.now()
datetime.datetime(2019, 8, 21, 12, 21, 34, 590280)
>>>import dateutil.parser
>>>dateutil.parser.parse("2019-06-03T18:18:21-0530")
>>>datetime.datetime(2019, 6, 3, 17, 30, 12, tzinfo=tzoffset(None, -18000))
```

Summary

- Any programming language requires making decision if the condition is true or false. If/else blocks are core syntax elements in Python to control the flow of program.
- Loops are used to repeat specific steps till a particular condition is true, then they stop automatically.
- Break and continue statements play an essential role in controlling the flow in a loop or any block.
- Using collections, we can store any data as a single variable. Some of them are mutable, and others are immutable.
- A big problem occurs when it involves time zone as time zones differ at various places. To manage this, we have specially designed date and time API.

Key terms

For controlling loops/blocks if, else, elif are used. Loops are of two types: while and for loop. Loops can be controlled by using break and continue statements. Collection can store immutable data as a single variable as per requirement. Date and time API is an excellent API to manage time zone.

Review questions

1. Which collection is anonymous locker?
2. What is the difference between list and tuples?
3. Can a float be a key in the dictionary?
4. Are duplicate keys allowed?
5. Is this a tuple ("Tarkeshwar")?
6. How would you stop while loop completely?

Exercise

Tick the correct option

Q.1. Which of these is not a core data type?
   a) List
   b) Dictionary
   c) Tuple
   d) Class

Q.2. Given a function that does not return any value, what value is thrown by default when executed in shell?
   a) int
   b) void
   c) bool
   d) None

Q.3. Which of the following runs without any error?
   a) round(45.8)
   b) round(6352.643,2,5)
   c) round()
   d) round(7463.234,2,2)

Q.4. What is the return type of function id?
    a) int
    b) float
    c) bool
    d) dict

Q.5. Which one of the following has the highest precedence in the expression?
    a) Exponential
    b) Addition
    c) Multiplication
    d) Parentheses

Q.6. Which is the correct operator for power(x,y)?
    a) X^y
    b) X**y
    c) X^^y
    d) None of the above

Q.7. What is the return value of trunc()?
    a) int
    b) bool
    c) float
    d) None

Q.8. What core data type is used in order to store value in terms of key and value?
    a) list
    b) tuple
    c) class
    d) dictionary

Q.9. What is the output of this expression 3*1**3?
    a) 9
    b) 3
    c) 27
    d) 1

Q.10. Which will be the output? 2 + 3*(2)
    a) 7
    b) 8
    c) 10
    d) 0.75

Answers

Q.1. d) Class
Q.2. d) None
Q.3. a) round(45.8)
Q.4. a) int
Q.5. d) Parentheses
Q.6. b) X**y
Q.7. a) int
Q.8. d) dictionary
Q.9. b) 3
Q.10. a) 8

Fill in the blanks

1. _____ statement helps you conditionally control the flow of program.
2. _____ operators have higher precedence than the logical operators like and/or/not.
3. _____ is a set of lockers.
4. Elements from the list can be accessible using _____.
5. _____ removes all duplicate elements automatically.

Answers

1. If else and elif
2. Comparison
3. Collection
4. index number
5. Set

4 Object-oriented programming (OOP)

> Most of you are familiar with the virtues of a programmer. There are three, of course: laziness, impatience, and hubris.
> ~Larry Wall

4.1 Object-oriented programming (OOP)

There are two types of programming approaches: procedural and object-oriented programming (OOP). In procedural programming, we write a program to handle data. This approach becomes very problematic while handling complex programs that require changes in content and scale. In OOP, however, we program object to handle data, and it is straightforward to make changes in content and scale at any moment in time. There are specific rules in OOP, they are as follows.

4.2 Class

Class is a blueprint of an object. It can be described as the map of a multistory building that has various attributes and behavior. It is an entity that has specific behaviors and attributes. Here, behavior is a method that is defined with **"self"** keyword. It refers to the instance of the same class that passes in a method as a parameter. The **"pass"** keyword refers to this, that the method block does nothing. We can prepare more than one building and with some of the changes apply to our customer demand. Likewise, in Python, we have a class that can have some attributes and behavior specifications. Let us see how to create a class in python:

```python
class Building(object):
    def __init__(self, flats, area):
        self.flats = flats
        self.area_in_yards = area
        super().__init__()

    def __str__(self) -> str:
        return super().__str__()

building1 = Building(20, 10000);
print("Building Ground Area", building1.area_in_yards)
print("Number of Flats in building - ", building1.flats)
building2 = Building(25, 12000);
print("Building Ground Area", building2.area_in_yards)
print("Number of Flats in building - ", building2.flats)
building3 = Building(30, 30000);
```

https://doi.org/10.1515/9783110689488-004

```
print("Building Ground Area", building3.area_in_yards)
print("Number of Flats in building - ", building3.flats)

-------------------------------output-------------------------------
Building Ground Area 10000
Number of Flats in building - 20
Building Ground Area 12000
Number of Flats in building - 25
Building Ground Area 30000
Number of Flats in building - 30
--------------------------------------------------------------------
```

In the above-mentioned example, we have seen that one class building is "like' a map and it can create many objects. Those objects will behave differently with their attributes.

4.3 Object

Python is an OOP language. An object contains its properties and methods and is created by a constructor of the class. Object can be created by calling the name of its class, which in turn calls the constructor. Constructor always returns an instance of the same class. In the given example, *p* is an object of class *Person*. A class can create many constructors and every object can have different properties and behavior:

```
>>> class Person:
. . . def __init__(self, name, phone):
. . .      self.name=name
. . .      self.phone=phone
. . .
>>>p=Person("Mr ABC", "3345564565")
>>>p.name
 'Mr ABC'
>>>p.phone
 '3345564565'
```

4.4 Functions

Function is stored as a class attribute. In the given example, get_color of the class Dog is a function, that is, unbound; therefore, we need to make it bound to an instance of Dog class. This is called function with an instance as its first argument, which is required to work with its attributes and behavior. In other programming languages, it is mandatory to create an object before using it by calling its constructor. Function needs it as the first argument as given in the solution:

```python
class Dog:
    def __init__(self, color):
        self.color=color

    def get_size(self):
        return self.color

print(Dog.get_color())
```

Output will be like this:

Traceback (most recent call last):
  File "/Users/admin/PycharmProjects/PythonBook/oop/methods.py", line 6, in
  <module> print(Dog.get_color())
TypeError: get_size() missing 1 required positional argument: "self"

Solution
```python
print(Dog.get_color(Dog("White"))
```
or
```python
dog=Dog("Black")
print(dog.get_color())
```
or
```python
dog1=Dog("black & White").get_color
print(dog1())
```

----------------------------Another example----------------------------
```python
class Dog:
    def __init__(self, color):
        self.color=color

    def get_color(self):
        return self.color

print(Dog.get_color(Dog("White")))
dog=Dog("Black")
print(dog.get_color())
dog1=Dog("black & White").get_color
print(dog1())
print(dog1.__name__)
```

```
print(dog1.__self__)
print("retrieving object of ", dog1.__self__.get_color)
print("Method is equal ", dog1==dog1.__self__.get_color)
```

```
--------------------------------output--------------------------------
White
Black
black & White
get_color
<__main__.Dog object at 0x000002640F5D0D48>
retriving object of <bound method Dog.get_color of <__main__.Dog object at
0x000002640F5D0D48>>
Method is equal True
----------------------------------------------------------------------
```

4.5 Abstraction

Abstraction means providing only essential information about the data to the users or outside world. It means hiding background details and implementation. Such an example is the air conditioner. It is fully covered so we only use a remote control to operate it. We are least concerned about its inner operations like how it cools and changes temperatures. Abstraction provides a blueprint for other classes. It allows us to create a set of methods that must be created within any child classes built from abstract class:

```python
class Shape:
    def countSides(self): #abstract method
        pass
class StraightLine(Shape):
    def countSides(self): #method overriding
        print("Straight Line have 1 sides")

class TwoStraightLine(Shape):
    def countSides(self): #method overriding
        print("Straight and parallel Line")

class Triangle(Shape):
    def countSides(self): #method overriding
        print("Triangle have 3 sides")

class Pentagon(Shape):
    def countSides(self): #method overriding
        print("Pentagone have 5 sides")
```

```
class Hexagon(Shape):
    def countSides(self): #method overriding
        print("Hexagon have 6 sides")

class Rectangle(Shape):
    def countSides(self): #method overriding
        print("Rectangle have 4 sides")

#main Program
straightLine=StraightLine()
straightLine.countSides()
twoStraightLine=TwoStraightLine()
twoStraightLine.countSides()
triangle = Triangle()
triangle.countSides()
rectangle = Rectangle()
rectangle.countSides()
pentagon = Pentagon()
pentagon.countSides()
hexagon = Hexagon()
hexagon.countSides()
```

```
----------------------------Output------------------------------------
Straight Line have 1 sides
Straight and parallel Line
Triangle have 3 sides
Rectangle have 4 sides
Pentagon have 5 sides
Hexagon have 6 sides
----------------------------------------------------------------------
```

4.5.1 Abstract class

It contains one or more abstract methods that are only declared, but their implementation can be a child class. In other words, abstract method is a method that is declared but contains no implementation. An abstract class cannot be instantiated and requires subclasses for implementation of its abstract methods:

```
def abstractmethod(method):
    def default_abstract_method(*args, **kwargs):
        raise NotImplementedError('abstract method call ' + repr(method))

    default_abstract_method.__name__ = method.__name__
```

```python
        return default_abstract_method
if __name__ == '__main__':
    class A:
        @abstractmethod
        def myAbstractMethod(self, data):
            pass

    class B(A):
        def myAbstractMethod(self, data):
            self.data = data
a = A()
b = B()
b.myAbstractMethod(20)
isItImplimented = False
try:
        b.myAbstractMethod(30)
        isItImplimented = False
        print("Class B, Method implementation is ", isItImplimented)

except NotImplementedError:
        isItImplimented = True
        print("Class B, Method implementation is ", isItImplimented)
try:
        a.myAbstractMethod(40)
        isItImplimented = False
        print("Class A, Method implementation is ", isItImplimented)

except NotImplementedError:
        isItImplimented = True
        print("Class A, Method implementation is ", isItImplimented)

-------------------------------output----------------------------------
Class B, Method implementation is False
Class A, Method implementation is True
------------------------------------------------------------------------
```

4.5.2 Interface

There is no interface keyword in Python that is why we can say that the interface is designed only for multiple implementations while Python supports multiple inheritances and duck typing (automatic interface). We are more focused on how an object behaves, rather than its type. Duck typing is a feature of a type system where the semantics of a class are determined by its ability to respond to some method or attributes, therefore, it doesn't make sense to implement the interface. "*We have an object that can fly and fly and quack like a duck, we can consider it as duck.*" Inheritance is very well implemented in Python. Still, there are so many usages of the interface. Interface was first introduced in Python 2.6. In the given example, dependency injection of color String class has been injected, it is not dependent on color; any type of color implementation can be injected in both classes. The following code is written in Java, which supports interface, but interfaces are not available in Python. We can compare many lines of code. If we are using Java, C, C++, C#, and Swift, then we can create an interface in Python by importing the interface library. For more information, we can visit https://pypi.org/project/python-interface/. It is required to be installed before using with pip command. The following example is explained first with using Java code and then the same code is written using python. Before writing code in python, ensure python interface library is installed as given by the following pip command:

```
$ pip install python-interface
--------------------------------Java Program--------------------------
  public interface Engine {
    void switchOn();
  }

  public class FourStrockEngine implements Engine {
    private String model, power;

    public FourStrockEngine(String model, String power) {
    this.model = model;
    this.power = power;
  }

  @Override
  public void switchOn() {
    System.out.println(model + " Four Strock Engine " + power);
  }
}
public class FourStrockTurboEngine implements Engine {
  private String model, power;

  public FourStrockTurboEngine(String model, String power) {
    this.model = model;
    this.power = power;
  }

  @Override
```

```java
    public void switchOn() {
      System.out.println(model + " Four Strock Turbo Engine " + power);
    }
}
public class Car {
  private Engine engine;

public Car(Engine engine) {
this.engine = engine;
  }
  public void run(){
    engine.switchOn();
  }
}
public class Test {
  public static void main(String[] args) {
    Engine fourStrockTurboEngine = new FourStrockTurboEngine("X1", "1000CC");
    Engine fourStrockEngine = new FourStrockEngine("Y1", "1500CC");
    Car neno = new Car(fourStrockTurboEngine);
    neno.run();
    Car audi = new Car(fourStrockEngine);
    audi.run();
  }
}
```

```
-------------------------------output--------------------------------
X1 Four Strock Turbo Engine 1000CC
Y1 Four Strock Engine 1500CC
----------------------------python program-------------------------
```

```python
>>>class Car:
. . . def __init__(self, Engine):
. . .     self.__Engine=Engine
. . . def run(self):
. . .     self.__Engine.switchOn()

>>>class Engine:
. . . def __init__(self, Model, Power):
. . .     self.__Model=Model
. . . self.__Power=Power

. . . def switchOn(self):
. . . print("Turning on Engine", self.__Model, " with power", self.__Power, "
has been started")
```

```
>>>fourStrockTurboEngine=Engine("Four Strock Turbo", "2000cc")
>>>fourStrockEngine=Engine("Four Strock", "1000cc")
>>>neno=Car(fourStrockEngine)
>>>neno.run()
'Turning on engine Four Strock with power 1000cc has been started'
>>>audi=Car(fourStrockTurboEngine)
>>>audi.run()
'Turning on engine Four Strock Turbo with power 2000cc has been started'
```

4.6 Polymorphism

Polymorphism is the ability to take various forms. It allows us to define methods in
the child classes with the same name as defined in their parent class. As we know,
during inheritance, child class shares the same state and behavior, if we make any
changes in parent class properties, it is called overriding:

```python
import math
class Shape:
    def __init__(self, color='black', filled=False):
        self.__color = color
        self.__filled = filled

    def get_color(self):
        return self.__color

    def set_color(self, color):
        self.__color = color

    def get_filled(self):
        return self.__filled

    def set_filled(self, filled):
        self.__filled = filled

    def is_filled(self):
        return self.__filled

class Rectangle(Shape):
    def __init__(self, length, breadth):
        super().__init__()
        self.__length = length
        self.__breadth = breadth
```

```python
    def get_length(self):
        return self.__length

    def set_length(self, length):
        self.__length = length

    def get_breadth(self):
        return self.__breadth

    def set_breadth(self, breadth):
        self.__breadth = breadth

    def get_area(self):
        return self.__length * self.__breadth

    def get_perimeter(self):
        return 2 * (self.__length + self.__breadth)

class Circle(Shape):
    def __init__(self, radius):
        super().__init__()
        self.__radius = radius

    def get_radius(self):
        return self.__radius

    def set_radius(self, radius):
        self.__radius = radius

    def get_area(self):
        return math.pi * self.__radius ** 2

    def get_perimeter(self):
        return 2 * math.pi * self.__radius

#Main Program
r1 = Rectangle(50, 5)
print("Area of rectangle r1:", r1.get_area())
print("Perimeter of rectangle r1:", r1.get_perimeter())
print("Color of rectangle r1:", r1.get_color())
print("Is rectangle r1 filled ? ", r1.get_filled())
r1.set_filled(True)
print("Is rectangle r1 filled ? ", r1.get_filled())
r1.set_color("orange")
print("Color of rectangle r1:", r1.get_color())
c1 = Circle(15)
print("\n Area of circle c1:", format(c1.get_area(), "0.2f"))
```

```
print("Perimeter of circle c1:", format(c1.get_perimeter(), "0.2f"))
print("Color of circle c1:", c1.get_color())
print("Is circle c1 filled ? ", c1.get_filled())
c1.set_filled(True)
print("Is circle c1 filled ? ", c1.get_filled())
c1.set_color("blue")
print("Color of circle c1:", c1.get_color())

-------------------------------output-------------------------------
Area of rectangle r1: 250
Perimeter of rectangle r1: 110
Color of rectangle r1: black
Is rectangle r1 filled ? False
Is rectangle r1 filled ? True
Color of rectangle r1: orange

Area of circle c1: 706.86
Perimeter of circle c1: 94.25
Color of circle c1: black
Is circle c1 filled ? False
Is circle c1 filled ? True
Color of circle c1: blue
-------------------------------------------------------------------
```

4.6.1 Overloading

The overloading feature does not exist in python, but something like this we can achieve using decorator **@classmethod.** Let us see it in the given code:

```
class Person:
    def __init__(self, name: str, address: str, phone: int):
        self.name = name
        self.address = address
        self.phone = phone

    @classmethod
    def classPersonConstructor(cls, person: str) ->'Person':
        return person

person=Person("Tarkeshwar", "Monrovia", "334423424")
secondObj = Person.classPersonConstructor(person)
print("Name ",secondObj.name)
print("Address ",secondObj.address)
```

```
print("Phone ",secondObj.phone)
```
---------------------------------output----------------------------
```
Name Tarkeshwar
Address Monrovia
Phone 334423424
```

And another approach is using default value for constructor function:

```
class Person:
    def __init__(self, name="No Name",address="No Address", phone="00000000"):
        self.name = name
        self.address = address
        self.phone = phone

    def __str__(self):
        return "Name : "+str(self.name)+" Address : "+str(self.address)+"
        Phone : "+ str(self.phone)

person1=Person()
person2=Person("Tarkeshwar")
person3=Person("Tarkeshwar", "Monrovia")
person4=Person("Tarkeshwar", "Monrovia", "334423424")
print("Person1 ",person1)
print("Person2 ",person2)
print("Person3 ",person3)
print("Person4 ",person4)
```

---------------------------output------------------------------------
```
Person1 Name : No Name Address : No Address Phone : 00000000
Person2 Name : Tarkeshwar Address : No Address Phone : 00000000
Person3 Name : Tarkeshwar Address : Monrovia Phone : 00000000
Person4 Name : Tarkeshwar Address : Monrovia Phone : 334423424
```

4.6.2 Overriding

Overriding is the feature of OOP. If we are creating more than one inherited classes in our program where two variables of same type and name exist, then child class can change its inherited property according to the given implementation inside child class. Python supports overriding. For example, Car and Bus are Vehicles; since both are sharing some common properties but max speed limit for both car and bus is different, it requires us to override according to the type of vehicle and it needs to be declared individually. The solution will be via override, which is to be written at runtime:

```python
class Vehicle(object):
    def __init__(self, model, max_speed):
        self.model = model
        self.max_speed = max_speed
        self.speed = 0

    def accelerate(self, speed_difference):
        self.speed += abs(speed_difference)
        self.speed = min(self.speed,self.max_speed)
    def slow_down(self, speed_difference):
        self.speed -= abs(speed_difference)
        self.speed = max(self.speed,-5)

    def __str__(self):
        return " "+self.model+" "+str(self.max_speed)
class Car(Vehicle):
    pass

class Bus(Vehicle):
    def slow_down(self, speed_difference): #Overriding parent class method
        super().slow_down(speed_difference)
        self.speed=max(self.speed,0)

bmw=Car("BMW-X6",300)
print(bmw)
bmw.max_speed=400
bmw.model="BMW-X7"
print(bmw.speed)
bmw.accelerate(30)
print(bmw.speed)
bmw.slow_down(5)
print(bmw.speed)
```

```
-------------------------------output------------------------------------
BMW-X6 300
0
30
25
--------------------------------------------------------------------------
```

4.7 Encapsulation

Encapsulation is the main and essential feature of OOP, building of data with the methods it operates on data. It is used to hide details of the inner workings of the system like values/state of the object so that the user does not need to know how it works in order to use/operate it. Python does not support it as strongly as Java, C, C++, Swift, and C#, the term is known as name mangling. It can prevent unauthorized access. The part of the program that the user interacts with is called "interface." The driver of a car needs tools such as a steering wheel and a gas pedal to use the capabilities of the car, in the same way users of our code require the syntax. They are not bothered about how the code works:

```python
class Car(object):
    def __init__(self, model, max_speed):
        self.model = model
        self.max_speed = max_speed
        self.speed = 0

    def accelerate(self, speed_difference):
        self.speed += abs(speed_difference)
        self.speed = min(self.speed, self.max_speed)

    def slow_down(self, speed_difference):
        self.speed -= abs(speed_difference)
        self.speed = max(self.speed, -5)

    def __str__(self):
        return " "+self.model+" "+str(self.max_speed)

bmw=Car("BMW-X6",300)
print(bmw)
bmw.max_speed=400
bmw.model="BMW-X7"
print(bmw.speed)

bmw.accelerate(30)

print(bmw.speed)
bmw.slow_down(5)
print(bmw.speed)
-------------------------------output-------------------------------
BMW-X6 300
0
30
25
--------------------------------------------------------------------
```

In the given code, object was created using two parameters: model and max speed. These parameters are passed through the constructor __init__ method and the values are copied as instance variable using self-keyword. After creation of object, values can be modified by using dot(.) notation. The constructor __init__ cannot have return statement because it always returns instance of the same class.

4.8 Inheritance

If we are creating more than one class in our program, then they must have a relationship, this is known as inheritance. Python supports multiple inheritances while other programming languages do not. For example, Car and Bus are Vehicles; since both share some common properties and need to be declared individually, the solution will be via inheritance, in which we can write once and read many:

```python
class Vehicle(object):
    def __init__(self, model, max_speed):
        self.model = model
        self.max_speed = max_speed
        self.speed = 0

    def accelerate(self, speed_difference):
        self.speed += abs(speed_difference)
        self.speed = min(self.speed,self.max_speed)

    def slow_down(self, speed_difference):
        self.speed -= abs(speed_difference)
        self.speed = max(self.speed,-5)

    def __str__(self):
        return " "+self.model+" "+str(self.max_speed)
class Car(Vehicle):
pass

class Bus(Vehicle):
    def slow_down(self, speed_difference): #Overriding parent class method
        super().slow_down(speed_difference)
        self.speed=max(self.speed,0)
bmw=Car("BMW-X6",300)
print(bmw)
bmw.max_speed=400
bmw.model="BMW-X7"
print(bmw.speed)
```

```
bmw.accelerate(30)
print(bmw.speed)
bmw.slow_down(5)
print(bmw.speed)

-------------------------------output--------------------------------
BMW-X6 300
0
30
25
---------------------------------------------------------------------
```

4.9 Access modifier/specifier

All members of the object are by default public! Python does not have an access modifier, but similar functionality can be achieved by using prefix underscore before any object and variable. Single underscore "_" stands for protected (module-level accessibility). It doesn't change its access permissions but shows "don't trust this part of the API to stay constant in behavior." It wouldn't be accessible while we use "**from moduleName import functionName**," in this case, we must define __all__ in that module and all those properties in it whatever we want to export in other modules. Double underscore "__" can be accessed by obj._ClassName__privateMethodName:

```
class SampleClass(object):
    def public_method(self):
        print("public method")

    def _single_underscore_method(self): # Protected method
        print("Protected/single-underscore method")

    def __double_underscore_method(self): # Private method
        print("private/double-underscore method")
obj1=SampleClass()
obj1.public_method()
obj2=SampleClass()
obj2._single_underscore_method()
obj3=SampleClass()
obj3.__double_underscore_method() # Can't be accessible due to private
privilege
```

```
--------------------------------output--------------------------------
Traceback (most recent call last):
File "C:/Users/Tarkeshwar/PycharmProjects/KiVyPrograms/OverloadingDemo.py",
line 13, in <module>
public method
obj3.__double_underscore_method() # Can't be accessible due to private privilege
AttributeError: 'SampleClass' object has no attribute '__double_underscore_method'
Protected/single-underscore method
----------------------------------------------------------------------
```

4.10 Decorator

In Python, everything is an object even classes. Decorators are very powerful and useful tools in Python, since it allows the programmer to modify the behavior of a function or a class. Functions and methods are called callable. Decorators are known as Meta programming too. A function takes another function and extends the behavior of the later function without explicitly modifying. Here, functions are taken as an argument into another function, then are called inside the wrapper function. It allows us to wrap another function in order to extend the behavior of the wrapped function without modifying permanently. Python may have side effects rather than just turning an input into an output. For example, print function, which returns None has the side effect in output something on the console. Functions are first-class objects (function can be passed around and used as arguments like other objects such as int, float, and string) as we can see in the given example:

```python
def greeting(msg):
  print(msg)

greeting("Hello, Good Morning")
second = greeting
second("Hello, Good Evening")
```

```
--------------------------------output--------------------------------
Hello, Good Morning
Hello, Good Evening
----------------------------------------------------------------------
```

Let us see another example by using annotation

```python
def smart_divide(func):
    def inner(a,b):
        print("Input values are ",a,"and",b)
```

```
        if b == 0:
            print("Wring Input")
            return
        return func(a,b)
  return inner

@smart_divide
def divide(a,b):
  return a/b

print(divide(10,20))
```

```
-------------------------------output-------------------------------
Input values are 10 and 20
0.5
--------------------------------------------------------------------
```

4.10.1 @classmethod decorator

We can create a class method that passes the actual class object within the function call like self-keyword. The self-argument is the class instance object itself, which can then be used to act on instance data but in the case of @classmethod, it is not the class instance. Let us see this example:

```
class Person(object):
  def __init__(self, first_name, last_name):
      self.first_name = first_name
      self.last_name = last_name

  def __str__(self) -> str:
      return super().__str__()

  @classmethod
  def split_string(cls, name_str):
      first_name, last_name = map(str, name_str.split(' '))
      student = cls(first_name, last_name)
      return student

testname=Person("Tarkeshwar"," Barua")
print(testname.first_name)
print(testname.last_name)
myname = Person.split_string('Tarkeshwar Barua')
```

```
print(myname.first_name)
print(myname.last_name)
```

```
-------------------------------output-------------------------------
Tarkeshwar
Barua
Tarkeshwar
Barua
--------------------------------------------------------------------
```

4.10.2 @staticmethod decorator

Like@*classmethod,* it can be called from an un-instantiated class object, it means we don't need to pass *cls* like class method decorator. Let us see the example:

```
class Person(object):
    @staticmethod
    def is_space_available(name_str):
        names = name_str.split(' ')
        return len(names) > 1
print("Available space is ",Person.is_space_available('Tarkeshwar Barua'))
print("Available space is ",Person.is_space_available('Tarkeshwar'))
```

```
-------------------------------output-------------------------------
Available space is True
Available space is False
--------------------------------------------------------------------
```

Summary

– A **Class** is a blueprint of an object. It specifies its attributes/state and behavior/ functions, for example, a Dog class can have attributes such as its legs, eyes, and color but its behavior or functions are – it can walk, bark, is friendly, aggressive, and so on.
– **Object** is a real entity created from the class. The object only occupies memory when it has been created.
– The constructor is responsible to create an object. The constructor by default created by the interpreter is called **"zero parameterized constructor."** Programmers can create their own parameterized constructors as well. After its creation, the interpreter will not create any constructor.

– **Polymorphism** can be achieved by using the abstract method and their implementation is done in the child class.
– **Abstraction** is code hiding. Implementation is carried out in child class; we are given access to its functionality not implementation.
– **Overriding and Overloading:** One object can behave in different ways in different situations. The two types of such behaviors are known as overriding and overloading. When a child class changes its property, which is inherited from the parent class, the term is known as overriding while overloading means that the same function can receive various parameters at runtime.
– **Encapsulation** is the ability to cover object attribute(s) and methods in a single entity with various access privileges.
– **Inheritance** is the capability to transfer parent class attributes and functions to the child class. Access specifier privileges to provide access to its attribute and methods.
– Overloading is not supported by Python, a similar type of functionality we can use to achieve this either by class method decorator or by default value feature of function.
– Decorators are a very powerful and useful tools in Python, since it allows the programmer to modify the behavior of function or class. Decorators are assigned by @ symbols in other programming languages.
– Decorators are known as annotations and these functions and methods are called callable as they can be called. Decorators are known as Meta programming too. Decorators are of two types: **staticmethod** and **classmethods.**

Key terms

Interfaces, Abstract classes, overloading, access specifier features are not supported by Python like Java, C, and C++, C#, to achieve these we have another similar feature.

Review questions

1. A collection of behavior that an user can use is called _____.

2. Relationship between various classes is called _____.

3. What is the use of constructor?

4. What does self-keyword do?

Exercise

Tick the correct option

Q.1.   Which function is override+operator?
a) __add__()
b) __plus__()
c) __sum__()
d) None of the above

Q.2.   Which of the following keyword is responsible to create class?
a) return
b) class
c) def
d) All of the above

Q.3.   class test: def __init__(self): print("Hello World") def __init__(self): print
("Bye World") obj=test()?
a) Bye World
b) Hello World
c) Compilation Error
d) Ambiguity

Q.4.   Which predefined Python function is used to find the length of string?
a) length()
b) len()
c) strlen()
d) stringlength()

Q.5.   Syntax of constructor in Python is _____
a) def __init__()
b) def _init_()
c) _init_()
d) All of the above

Q.6.   Get the last element of list named with names:
a) names[0]
b) names[-1]
c) names[|pos]
d) names[:-1]

Q.7. Mention the method that begins and ends with two underscore characters.
a) Special method
b) In-built method
c) User-defined method
d) Additional method

Q.8. Which of the following is not OOP concept in Python?
a) Inheritance
b) Encapsulation
c) Polymorphism
d) Compilation

Q.9. Which concept of Python is a way of converting real-world objects in terms of class?
a) Polymorphism
b) Encapsulation
c) Abstraction
d) Inheritance

Q.10. Which concept of Python is achieved by combining methods and attributes in a class?
a) Encapsulation
b) Inheritance
c) Polymorphism
d) Abstraction

Answers

Q.1. a) __add__()
Q.2. b) class
Q.3. a) Bye World
Q.4. b) len()
Q.5. a) def __init__()
Q.6. b) names[-1]
Q.7. a) Special method
Q.8. d) Compilation
Q.9. c) Abstraction
Q.10. a) Encapsulation

Fill in the blanks

1. Class is a _____of an object.
2. One or more abstract methods are only declared but their _____can be a child class.
3. Object contains its_____.
4. _____allow programmers to modify the behavior of function or class.
5. Python supports _____while other programming languages do not.

Answers

1. blueprint
2. implementation
3. properties and methods
4. Decorators
5. multiple inheritance

5 Standard libraries

If you're talking about Java in particular, Python is about the best fit you can get among all the other languages. Yet the funny thing is, from a language point of view, JavaScript has a lot in common with Python, but it is sort of restricted subset.   ~Guido van Rossum

5.1 File operations using native

Programmers are needed to open, read, write, and close files. These files are mostly written in ANSI textual (UTF-8) format. Files can be opened using the built-in function open. Using <command> as <name> syntax makes using open and getting a handle for the file super easy. One file can be opened in different modes; here we are opening in read only mode. The first argument to the "open" function is the path to the file, and the second argument is related to the file name where we want to write. Open function returns an object of file using which, we can write data into the given file. The write method takes a string to write into the file. After doing all this, we require to close file object, which discontinues and deletes connection from the file. It resides in the main memory. If we are willing to read any file, then we do not need any permission. Read permission is given by default. After getting a file object from the open function, we can use *readline()* or *readlines()* to read single or multiple lines at one time; later resulting objects can be retrieved or iterated from the given object. After doing all these, task files are required to be closed using function *close()*:

```
>>>with open('fileName.txt', 'r') as fileReadObj        #read only
>>>fileReadObj.readlines()                              # reading all the line in one time
>>>fileReadObj
>>>with open('fileName.txt', 'rb') as fileReadObj       #byte use read only
>>>with open('fileName.txt', 'rb+') as fileReadObj      #byte use read and write
>>>with open('fileName.txt', 'a') as fileReadObj        #append mode
>>>with open('fileName.txt', 'a+') as fileReadObj       #appending reading mode
>>>with open('fileName.txt', 'ab') as fileReadObj       #append in binary mode
>>>with open('fileName.txt', 'ab') as fileReadObj       #append reading in binary
mode
>>>with open('fileName.txt', 'w') as fileReadObj        #overwrite write mode
>>>with open('fileName.txt', 'w+') as fileReadObj       #overwrite write and read
mode
>>>with open('fileName.txt', 'wb+') as fileReadObj      #overwrite write and read
in binary mode
>>>fileReadObject.write("Writing line in to file")      # writing a line into file
>>>with open('fileName.txt') as fileReadObj             #default is read only
```

https://doi.org/10.1515/9783110689488-005

```
>>>with fileReadObj.read()            #Read file as one line
>>>with open('fileName.txt', 'r+') as fileReadObj     #read and write mode
```

5.1.1 Writing file

Writing files in Python is very easy; **open** is a default method to write the file. Python does not automatically add line breaks; we need to use **\n** and **\t** for tab. Encoding parameters can be added along with open function. In case the file does not exist, then it will be created:

```
>>>with open('hello.txt', 'w', encoding='utf-8') as txt:
. . . txt.write("Hello World\n")
. . . txt.write("This is Second Line")
     print("This Text Using print function", file=txt)


-------------------------Example by Pycharm-------------------------
txt=open("hello.txt","w")
txt.write("First Line in the File\n")
txt.write("Second and lastLine in the File")
txt.close()
```

5.2 OS module

Built-in python functions are enough to work with files, but still, they have specific problems that we may face in the manipulation of files. Let us take an example, we have developed a program that works with Windows but the same program does not work in Linux because of / (forward slash used in Linux and Mac) and \ (backward slash used in Windows). OS module can resolve it very efficiently. To get the absolute path of the current directory, functions such as **abspath** and **curdir** from the os.path module can be used. Let us have a look over the given code:

```
>>>from os.path import abspath, curdir
>>>curdir
'.'
>>>abspath(curdir)
'/parent'
>>>dir(os)
```

['DirEntry', 'F_OK', 'MutableMapping', 'O_APPEND', 'O_BINARY', 'O_CREAT', 'O_EXCL', 'O_NOINHERIT', 'O_RANDOM', 'O_RDONLY', 'O_RDWR', 'O_SEQUENTIAL', 'O_SHORT_LIVED', 'O_TEMPORARY', 'O_TEXT', 'O_TRUNC', 'O_WRONLY', 'P_DETACH', 'P_NOWAIT', 'P_NOWAITO', 'P_OVERLAY', 'P_WAIT', 'PathLike', 'R_OK', 'SEEK_CUR', 'SEEK_END', 'SEEK_SET', 'TMP_MAX', 'W_OK', 'X_OK', '_Environ', '__all__', '__builtins__', '__cached__', '__doc__', '__file__', '__loader__', '__name__', '__package__', '__spec__', '_execvpe', '_exists', '_exit', '_fspath', '_get_exports_list', '_putenv', '_unsetenv', '_wrap_close', 'abc', 'abort', 'access', 'altsep', 'chdir', 'chmod', 'close', 'closerange', 'cpu_count', 'curdir', 'defpath', 'device_encoding', 'devnull', 'dup', 'dup2', 'environ', 'error', 'execl', 'execle', 'execlp', 'execlpe', 'execv', 'execve', 'execvp', 'execvpe', 'extsep', 'fdopen', 'fsdecode', 'fsencode', 'fspath', 'fstat', 'fsync', 'ftruncate', 'get_exec_path', 'get_handle_inheritable', 'get_inheritable', 'get_terminal_size', 'getcwd', 'getcwdb', 'getenv', 'getlogin', 'getpid', 'getppid', 'isatty', 'kill', 'linesep', 'link', 'listdir', 'lseek', 'lstat', 'makedirs', 'mkdir', 'name', 'open', 'pardir', 'path', 'pathsep', 'pipe', 'popen', 'putenv', 'read', 'readlink', 'remove', 'removedirs', 'rename', 'renames', 'replace', 'rmdir', 'scandir', 'sep', 'set_handle_inheritable', 'set_inheritable', 'spawnl', 'spawnle', 'spawnv', 'spawnve', 'st', 'startfile', 'stat', 'stat_result', 'statvfs_result', 'strerror', 'supports_bytes_environ', 'supports_dir_fd', 'supports_effective_ids', 'supports_fd', 'supports_follow_symlinks', 'symlink', 'sys', 'system', 'terminal_size', 'times', 'times_result', 'truncate', 'umask', 'uname_result', 'unlink', 'urandom', 'utime', 'waitpid', 'walk', 'write']

>>>os.__doc__

"OS routines for NT or Posix depending on what system we're on.\n\nThis exports: \n - all functions from posix or nt, e.g. unlink, stat, etc.\n - os.path is either posixpath or ntpath\n - os.name is either 'posix' or 'nt'\n - os.curdir is a string representing the current directory (always '.')\n - os.pardir is a string representing the parent directory (always '..')\n - os.sep is the (or a most common) pathname separator ('/' or '\\\\')\n - os.extsep is the extension separator (always '.')\n - os.altsep is the alternate pathname separator (None or '/')\n - os.pathsep is the component separator used in $PATH etc\n - os.linesep is the line separator in text files ('\\r' or '\\n' or '\\r\\n')\n - os.defpath is the default search path for executables\ - os.devnull is the file path of the null device ('/dev/null', etc.)\\Programs that import and use 'os' stand a better chance of being\nportable between different platforms. Of course, they must then \nonly use functions that are defined by all platforms (e.g., unlink\nand opendir), and leave all pathname manipulation to os.path\n(e.g., split and join).\n"

Curdir variable holds a string pointing to the current directory, while abspath expects any path as its argument and returns the absolute path. We can do create, read, update, and delete (CRUD) operations over files in any operating system. We are required to remove and rename functions from the OS module:

```
>>>from os import remove, rename
>>>remove("test/one.txt")
>>>rename("test/two.txt", "test/three.txt")
```

During any operation, we have to pass the file name, either rename or remove the file. File path can be absolute or relative, in case, an absolute path is plain; it starts with the root of the file systems while relative paths are calculated with the current directory. Before the deletion of any directory, it must be empty; otherwise, it will raise an exception:

```
#/home/tarkeshwar/Desktop/test/one.txt file want delete
#C:\Users\tarkeshwar\Desktop\test\one.txt file want to rename to three,txt
>>>move, rename, listdir, rmdir
>>>from os.path import join
>>>remover(join("test", "one.txt"))
>>>rename(join("test", "two.txt"), join("text", "three.txt"))
>>>listdir("test")
["three.txt", "four.txt"]
>>>rmdir("test")
Traceback (most recent call last):
    File "<stdin>", line 1, in <module>
OSError:[Errno 39] Directory no empty: 'test'
```

5.3 Exception handling

Sometimes it happens that we are trying to access some file, but unfortunately, that file was deleted or does not exist. In this case, it may raise some exceptions, and the program execution stops. In another scenario like we are willing to divide 10/0, which is impossible, it will raise an exception. This is called exception traceback. This situation can be handled using try and catch statement. In the raised exception, the last line tells us which exception was raised and gives us some additional, human-readable information:

```
>>>10/0
Traceback (most recent call last):
    File "<stdin>", line 1, in <module>
ZeroDivisionError: division by zero
>>>try:
```

```
. . . 10/0
. . .except ZeroDivisionError as e:
. . . print("Error Caught in the try block")
```

Our code should be placed inside the try block, and exception handling code comes under except block, except keyword must be followed by a respective class of exception. Some alias name or variable name can be provided to this exception class. This variable will be used to store a human-readable description of the error.

5.3.1 User-defined exceptions

We can define our exceptions too; we need to inherit that particular class to whom it is related or parent exception class:

```
>>>class MyCustomException(Exception):
. . . pass
>>>try:
. . . raise MyCustomException("This this custom Raised Exception")
  Except MyCustomException as e:
. . . print("Error Caught")
. . .
Error Caught
```

5.4 Import standard Python modules and packages

Python 2 provides functionality to import from python 3 using __future__ module. If we are using Python 2 then the following statement will be used to print statements:

```
        print 'Hello World'      #python 2
Or
        print "Hello World"      #python 2
Or
        print('Hello World')      #python 3
Or
        print ("Hello world")     # python 3
```

The package can import complete, partial, and some methods along with an alias name. A module can be a stand-alone runnable script. If the module is inside a directory and needs to be detected by Python, the directory should contain a file named __init__.py:

```
#main_app.py
if __name__ == '__main__':
        from hello import sayHello
        sayHello()
```

5.4.1 Creating a module

A module is an importable file containing definitions and statements. A module can be created by creating a **.py** file. These files are called modules. Different modules are logically combined together into collections called packages:

```
#save this file as hello.py in same directory
def sayHello():                        #file name hello.py in same directory
        print("Hello World")
>>>import hello
>>>hello.sayHello()
Hello World
>>>dir(hello)
['__annotations__', '__call__', '__class__', '__closure__', '__code__',
'__defaults__', '__delattr__', '__dict__', '__dir__', '__doc__', '__eq__',
'__format__', '__ge__', '__get__', '__getattribute__', '__globals__',
'__gt__', '__hash__', '__init__', '__init_subclass__', '__kwdefaults__',
'__le__', '__lt__', '__module__', '__name__', '__ne__', '__new__', '__qual-
name__', '__reduce__', '__reduce_ex__', '__repr__', '__setattr__', '__si-
zeof__', '__str__', '__subclasshook__']
```

5.4.2 Packages

A package is a plain folder with **__init__.py** file. Package is the way of managing code easily to develop application. Default Python installation comes with a significant number of packages. This set of packages is called standard libraries. Here, import statement plays a very important role to provide access of built-in and user-defined package attributes.

As we can see project name is Python book, under it we have one python package named with **test_package.** Inside it, we wrote two modules Calculator .py and Calculator1.py. Their codes are given as follows:

```
class Calc: #module name Calculator.py
    def addition(self, a, b):
        return a+b;
```

```
    def subtraction(self, a, b):
        return a-b;
    def multiplication(self, a, b):
        return a*b;
    def devision(self, a, b):
        return a/b;
    def modulo(self, a, b):
        return a%b;
```

-----------------------------another module--------------------------

```
def addition(a, b): #module name Calculator1.py
    return a + b;
def subtraction(a, b):
    return a - b;
def multiplication(a, b):
    return a * b;
def devision(a, b):
    return a / b;
def modulo(a, b):
    return a % b;
```

Testing and main code to import entire class attributes
```
from test_package.Calculator import Calc
print(Calc().addition(2,4))
print(Calc().subtraction(2,4))
print(Calc().multiplication(2,4))
print(Calc().devision(2,4))
print(Calc().modulo(2,4))
```

Testing and main code to import specific feature from the module
```
from test_package.Calculator1 import addition, multiplication, subtraction,
division, modulo
print(addition(2,2))
print(multiplication(2,2))
print(subtraction(2,2))
print(devision(2,2))
print(modulo(2,2))
```

In the above-mentioned code, it is shown how to import the most top-level package. As we can see, members of the package are supposed to be accessed using Dot operator. We may avoid Dot operator by importing the things directly, as displayed in second code. To indicate the source of the import we must use "**from**" keyword. In this case, it is very simple to import all entities directly. These entities, kept

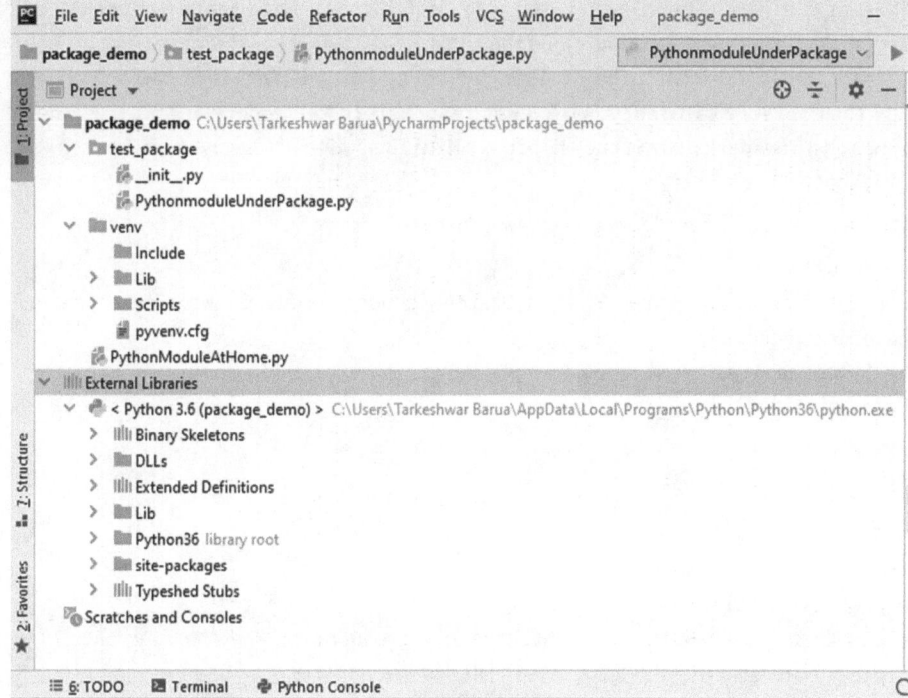

Figure 5.1: Python package.

inside python packages, can be virtually any Python constructs, class, function, variable module, or even package. Package can be nested.

5.5 Regular expression

Regular expression is used to search for any substring based on some specific conditions. Anyway, we have in operator that can do the same thing but it doesn't tell us where the substring is located and this becomes a very effort-consuming job for the computer when it comes to search from a big string. In the given string we are willing to search like "concern . . . . .war." Search function takes a regular expression argument and the second argument in which we want to search a string. The result is returned as MatchObject, which contains all the information about the result. In case multiple occurrences are found, then only the first one will be returned. If we want to match any string re.match() function is there where * matches to zero or more characters while. dot represents a single character. Let us see the given problem.

```
>>>test_string= "The Ministry of Truth, which concerned itself with news, en-
tertainment, education and the fine arts. The Ministry of Peace, which con-
cerned itself with war. The Ministry of Love, which maintained law and order.
And the Ministry of Plenty, which was responsible for economic affairs. Their
names, in Newspeak: Minitrue, Minipax, Miniluv and Miniplenty."
>>>import re
```

```
>>>result=re.search("concerned.*war", test_string)
```

```
>>>result
<re.Match object; span=(29, 155), match='concerned itself with news, enter-
tainment, educat>
>>>type(result)
<class 're.Match'>
```

```
>>>result.start()
29
```

```
>>>result.end()
155
```

```
>>>result.group()
'concerned itself with news, entertainment, education and the fine arts. The
Ministry of Peace, which concerned itself with war'
```

```
>>>test_string[29:155]
'concerned itself with news, entertainment, education and the fine arts. The
Ministry of Peace, which concerned itself with war'
```

```
>>>result1=re.finditer("Ministry", test_string)
```

```
>>>for data in result1:
...    print(data)
...
<re.Match object; span=(4, 12), match= 'Ministry'>
<re.Match object; span=(105, 113), match= 'Ministry'>
<re.Match object; span=(161, 169), match= 'Ministry'>
<re.Match object; span=(219, 227), match= 'Ministry'>
```

```
>>>bool(re.match("[0-9][a-z]*", "8dfdjhfjdf"))
True
```

```
>>>bool(re.match("[0-9][a-z]*", "fdjhfjdf"))
False
```

```
>>>dir(re)
```

```
['A', 'ASCII', 'DEBUG', 'DOTALL', 'I', 'IGNORECASE', 'L', 'LOCALE', 'M',
'MULTILINE', 'Match', 'Pattern', 'RegexFlag', 'S', 'Scanner', 'T', 'TEMPLATE',
'U', 'UNICODE', 'VERBOSE', 'X', '_MAXCACHE', '__all__', '__builtins__',
'__cached__', '__doc__', '__file__', '__loader__', '__name__', '__package__',
```

```
'__spec__', '__version__', '_cache', '_compile', '_compile_repl', '_expand',
'_locale', '_pickle', '_special_chars_map', '_subx', 'compile', 'copyreg',
'enum', 'error', 'escape', 'findall', 'finditer', 'fullmatch', 'functools',
'match', 'purge', 'search', 'split', 'sre_compile', 'sre_parse', 'sub', 'subn',
'template']
```

Frequently used cardinalities are given in the list (Table 5.1).

Table 5.1: List of wild cards.

Cardinality	Description
*	Zero or more
+	One or more
?	Zero or one
{m}	Exact number of symbols
{m, n}	Minimum and maximum number
^	Matches the beginning of a line also can mean "NOT"
$	Matches the end of a line
[a-zA-Z0-9_]	Any letter small and caps, any number, underscore
[^a-z]	Any character that is not a lower letter
\n \t \\ \' \"	new line character, tab, slash, single quote, double quote

5.6 Math module

Python has a math module that provides most of the familiar mathematical functions. Module is a file that contains a collection of related functions and variables defined. To access one of the functions, we have to specify the name of the module, which is called the name of the function separated by dot(.), Built-in modules contain some extra functionalities. __doc__ is useful to provide some documentation:

```
>>> import math
>>> math.sin(30)*math.sin(30)+math.cos(30)*math.cos(30)
1.0
>>>math.sqrt(3)
1.7320508075688772
>>>math.sin(30)
```

```
0.9880316240928618
>>>r=4.56767
>>>math.pi*r*r
65.54496148088045
>>>dir(math)
'__doc__', '__file__', '__loader__', '__name__', '__package__', '__spec__',
'acos', 'acosh', 'asin', 'asinh', 'atan', 'atan2', 'atanh', 'ceil', 'copy-
sign', 'cos', 'cosh', 'degrees', 'e', 'erf', 'erfc', 'exp', 'expm1', 'fabs',
'factorial', 'floor', 'fmod', 'frexp', 'fsum', 'gamma', 'gcd', 'hypot',
'inf', 'isclose', 'isfinite', 'isinf', 'isnan', 'ldexp', 'lgamma', 'log',
'log10', 'log1p', 'log2', 'modf', 'nan', 'pi', 'pow', 'radians', 'remain-
der', 'sin', 'sinh', 'sqrt', 'tan', 'tanh', 'tau', 'trunc']
>>>math.__doc__
'This module is always available. It provides access to the mathematical func-
tions defined by the C standard.'
>>>math.__file__
'/anaconda3/lib/python3.7/lib-dynload/math.cpython-37m-darwin.so'
>>>math.__loader__
<_frozen_importlib_external.ExtensionFileLoader object at 0x10a1bfb00>
>>>math.__name__
'math'
>>>math.__package__
''
>>>math.__spac__
ModuleSpec(name='math', loader=<_frozen_importlib_external.ExtensionFileLoader
object at 0x10a1bfb00>, origin='/anaconda3/lib/python3.7/lib-dynload/math.cpy-
thon-37m-darwin.so')
```

5.6.1 Writing documentation for module

Documentation is a very important part of our module. It provides complete information. Sample module documentation is given as follows:

```
""" This is the module doc string """
def sayHello():
      """This is the function doc string """
      return "Hello World"
>>>sayHello()
```

```
>>>import testmodule

>>>testmodule.__doc__
This is the module doc string

>>>testmodule.sayHello.__doc__
This is the function doc string

>>>dir(testmodule)
['__builtins__', '__cached__', '__doc__', '__file__', '__loader__',
'__name__', '__package__', '__spec__', 'sayHello']
```

5.7 JSON

JSON stands for JavaScript Object Notation. In the process of serialization/deserialization, it becomes very important to pass JSON as input and output. JSON is a string with key and value pair. It is a string that is readable by human and machine easily. Converting from objects to JSON is called serialization while its opposite from JSON to objects is known as deserialization. Data can be transmitted over the network in two forms; one is JSON, and the second is XML. Both transfer date in text format. JSON is the primary mechanism to handle date in JavaScript programming language. JSON can handle the following types of data type:
- **Primitive:** integer, string, float, boolean, null
- **Object:** a group of primitive variables with key-value pair
- **Arrays:** a group of primitive values

By default, python can serialize int, float, long, dict, list, str, bool, None. Python provides JSON module to handle serialization and deserialization. It has an important function-dumps and loads. A dump expects a python date structure as an input and returns JSON string. A load does just the opposite of it. We pass JSON string as input and output will be a Python object. We can have a look on the given code:

```
>>student={'name': 'Mr. ABC', 'age':25, 'height': 5.6, 'isGraduate': True,
'marks':None}

>>>import json

>>>type(student)
<class 'dict'>

>>>json_data=json.dumps(student)

>>>type(json_data)
<class 'str'>

>>>json_data
```

'{"name": "Mr. ABC", "age": 25, "height": 5.6, "isGraduate": true, "marks": null}'

Let us see how to convert from JSON string to python object:

```
>>>import json
```

```
>>>json_str='{"name": "Mr. ABC", "age": 25, "height": 5.6, "isGraduate": true, "marks": null}'
```

```
>>>type(json_str)
<class 'str'>
```

```
>>>python_obj=json.loads(json_str)
```

```
>>>type(python_obj)
<class 'dict'>
```

```
>>>python_obj
{'name': 'Mr. ABC', 'age':25, 'height': 5.6, 'isGraduate': True, 'marks': None}
```

```
>>>dir(json)
['JSONDecodeError', 'JSONDecoder', 'JSONEncoder', '__all__', '__author__', '__builtins__', '__cached__', '__doc__', '__file__', '__loader__', '__name__', '__package__', '__path__', '__spec__', '__version__', '_default_decoder', '_default_encoder', 'codecs', 'decoder', 'detect_encoding', 'dump', 'dumps', 'encoder', 'load', 'loads', 'scanner']
```

Like this, any python object can be converted to JSON string and vice versa. In case any problem is found during conversion, it would raise **JSONDecodeError, NameError** as we can see in the given code:

```
>>>import json
```

```
>>>json_str='Mr. ABC'
```

```
>>>type(json_str)
<class 'str'>
```

```
>>>python_obj=json.loads(json_str)
Traceback (most recent call last):
    File "<stdin>", line 1, in <module>
. . .. .. . .. ..
Json.decoder.JSONDecodeError: Expecting value: line 1 column 1 (char 0)
```

```
>>>json.dumps(Mr. ABC)
Traceback (most recent call last):
    File "<stdin>", line 1, in <module>
NameError: name 'Mr. ABC' is not defined
```

5.8 SQLite3

Every programmer wants to save data in the server for further use. To save data over a server, it requires getting one Relational Database Management System (RDBMS) server such as SQLite, MySQL, MSSQL Server, and Maria DB, and to install these servers it requires more resources. This server may be available on the same machine (localhost), remote machine (Internet/Intranet), and local file too. It is always a good approach to test our application with the SQLite database. It is file-based and open-source database. It does not require any type of server installation. SQLite database is very easy to share with others and no special SQL client is required. One of the main principles of software development is re-usability and Do Not Repeat Yourself (DRY). Before storing information in database we have to define the structure of our data in the form of a table. Information processing and data storage are achieved by some common set of constraints (set of rules) that have emerged. A relational database is made up of tables, rows, and columns.

RDBMS is based on relating information with each other in the form of tables where each column represents some set of rules to store data in a respective column such as int, char, float, and boolean. Primary key constrains define that no data can be duplicated. A cursor is like a file handle we can use to perform operations on the data stored in the database. Calling cursor() is like opening a text file where we can perform other operations such as CRUD over any database. Index means some unique identity number in sorting order that makes searching information easy in the large table. SQL stands for Structured Query Language, which is used to do so in RDBMS. It is a programming language for SQL constraint programming.

For example, one person can have zero or many phone numbers. This type of information can be related using foreign key constraints. Python comes with a built-in sqlite3 library. It is a full-stack implementation of RDBMS. Let us see how to handle the database in Python. SQLite version 3.6.x or above supports foreign key constraints too. By default, foreign key is deactivated, we can activate it with "PRAGMA foreign_key=ON;". Cursor is a python object to execute the SQL statements. In the given code on delete cascade tells SQL when an employee deletes first, all related cars must be deleted:

```
>>>import sqlite3
>>>conn=sqlite3.connect("sampledb.sqlite")
>>>cursor=conn.cursor()
>>>cursor.execute("PRAGMA foreign_keys= ON")
>>>cursor.execute("create table Employee(id integer not null primary key
autoincrement, name text not null, salary numeric not null);")
<sqlite2.Cursor object at 0x107f28ab0>
```

```
>>>cursor.execute("insert into Employee values(1, 'Mr. ABC', 100);")
<sqlite2.Cursor object at 0x107f28ab0>
>>>cursor.execute("insert into Employee values(1, 'Mr. ABC', 100);")
Traceback (most recent call last):
    File "<stdin>", line 1, in <module>
Sqlite3.IntegrityError: UNIQUE constraint failed: Employee.id
>>>cursor.execute("insert into Employee(name, salary) values('Mr. KBC', 150);")
<sqlite2.Cursor object at 0x107f28ab0>
>>>cursor.execute("insert into Employee values(3, 'Mr. XYZ', 175);")
<sqlite2.Cursor object at 0x107f28ab0>
>>conn.commit()
>>>result=cursor.execute("select * from Employee;")
>>>for record in result:
. . . print(record)
. . .
(1, 'Mr. ABC', 100)
(2, 'Mr. KBC', 150)
(3, 'Mr. XYZ', 175)
>>>cursor.execute("update Employee set salary=200 where id=1;")
>>>result=cursor.execute("select * from Employee;")
>>>for record in result:
. . .    print(record)
. . .
(1, 'Mr. ABC', 200)
(2, 'Mr. KBC', 150)
(3, 'Mr. XYZ', 175)
>>>cursor.execute("create table Car(id integer not null primary key auto-
increment, brand text not null, model text not null, owner integer not null
references Employee on delete cascade);")
<sqlite2.Cursor object at 0x107f28ab0>
>>>cursor.execute("insert into Car values(1, 'Maruti', 'Maruti 800', 1);")
<sqlite2.Cursor object at 0x107f28ab0>
>>>cursor.execute("insert into Car values(1, 'Maruti', 'Swift', (select id
from Employee where name= 'Mr. XYZ'));")
<sqlite2.Cursor object at 0x107f28ab0>
>>>cursor.execute("insert into Employee values(1, 'Mr. ABC', 100);")
Traceback (most recent call last):
    File "<stdin>", line 1, in <module>
Sqlite3.IntegrityError: UNIQUE constraint failed: Employee.id
```

```
>>>cursor.execute("insert into Car values(1, 'Maruti', 'Maruti 800', 1);")
Traceback (most recent call last):
File "<stdin>", line 1, in <module>
Sqlite3.IntegrityError: UNIQUE constraint failed: Car.id
```

5.9 SQLAlchemy

SQLAlchemy is an object-relational mapping (ORM) tool to map database tables with python objects, created by Mike Bayer. It provides a database server-independent SQL expression language, no more SQL injection, no more complex SQL queries, working with objects not tables, and fast execution of queries. It uses a data mapper pattern. The following are the four significant components:

–   **Engine** – connection pool management and database-independent SQL dialect layer.
–   **MetaData** – managing information about our database tables.
–   **SQL expression language** – to execute SQL queries and update against our table.
–   **ORM** – database persistence to our Python objects without requiring us to design around database.

```
>>>import sqlalchemy
>>>sqlalchemy.__version__
'1.2.15'
>>>engine=sqlalchemy.create_engine("sqlite:///testdb.sqlite", echo=True)

>>>engine=sqlalchemy.create_engine("postgresql://root:password@local-
host/testdb", echo=True)

>>>engine=sqlalchemy.create_engine("
mysql+pymysql://root:password@localhost/testdb", echo=True)

>>>metadata=sqlalchemy.MetaData()

>>>persons_table=sqlalchemy.Table('persons', metadata, sqlalchemy.Column
('id', sqlalchemy.Integer, primary_key=True), sqlalchemy.Column('name',
sqlalchemy.String))

>>>phone=sqlalchemy.Table('phone',  metadata,sqlalchemy.Column('id',  sqlal-
chemy.Integer,   primary_key=True),    sqlalchemy.Column('number',sqlalchemy.
String), sqlalchemy.Column('person_id', sqlalchemy.ForeignKey('persons.id')))

>>>metdata.create_all(engine)

>>>test=persons_table.insert(bind=engine)
```

```
>>>test.execute(name="Tarkeshwar")
>>>del_stmt=persons_table.delete()
>>>del_stmt.execute(name= "Tarkeshwar")
```

Summary

The highlights of this chapter are as follows:
- How to do basic file operations like read–write, and open file using built and standard libraries quickly?
- Performing basic tasks like creating, deleting, renaming files and folders by OS module.
- File and folder management such as current path of file and folder to perform some task, from given file or directory.
- All the created files, to write Python code into a file known as a module, and how to import modules in our package.
- Exception handling and creating user-defined exceptions using Exception class.
- Creating packages apart from built-in packages and importing packages partially or entirely.
- Regular expression to find and replace any string on the basis of the given wild cards.
- Math module to perform all the mathematical calculations efficiently.
- JSON is commonly used to transfer data between machines.
- SQLite is the small file-based database that supports all the RDBMS standards. Being a file-based database, files can be transferred easily over the poor networks too.
- SQLAlchemy is one of the most popular object-relational Mapper (ORM) tool. Using it we can generate complex SQL queries to interact with any database by handling Python objects.

Key terms

Python supports write once and read many. To achieve this, the user can perform file operations by OS module, creating package and modules and then importing them into the useful writing code in the module, and the method to access module in our code. During such operations, we may face unexpected problems like we are trying to access the file but unfortunately that is not available in the system problem. This is called exception and this may cause an interruption in the program during execution. This type of situation can be handled by exception handling. Exceptions are of two types – checked (compile time) and unchecked (runtime exception). User can create their own exception classes to handle. Regular expression is searching based on the wildcard based on the given string. Math module is used to perform the mathematical calculations as per requirement. Mostly all the mathematical formulas are built-in available. JSON is flexible and easy to transmit information over the network. JSON information is readable by humans and machines. It supports all the available devices. This is one of the leading causes of its popularity among programmers. Organizations have always been storing and managing business data using RDBMS, which requires to install and manage server that needs a lot of resources and efforts. We can achieve this by using very lightweight databases SQLite. It is file-based, cross-platform, open-source RDBMS. We can use it for testing purposes. It supports all the features of RDBMS. SQL queries writing is a challenging task, and it requires to learn SQL ueries individually now, we have a solution one of the ORM tool SQLAlchemy. By using the SQLAlchemy, the programmer can focus on writing code instead of SQL ueries.

Review questions

1. What is the synonym of the Python source file?
2. How do we import module path from OS package?
3. What is the significant difference between Windows and Linux file systems to make it cross-platform?
4. What is the value of os.path.curdir?
5. Write a code to remove a nonempty directory.
6. Write a program to list all the files and directory from the current directory.
7. What is the escape sequence for the new line and tab character?
8. What is the use of re.search() function?
9. What is responsible for JSON serialization in Python?
10. What will be the output of the command: **print(json.dumps("Mr. KBC")[0])**?
11. What will be the output of the command: **print(type(json.dumps("Mr. KBC")))**?
12. How would you join two arrays in JSON?
13. What are the meanings of DRY and SQL?
14. In what form is data stored inside the relational database?
15. Constraints are usually applied to which part of a database?
16. Write a program to interact with the database and perform CRUD (create, read, update, delete) operations using SQLAlchemy.

Exercise

Tick the correct option

Q.1. Which of the following function is a built-in function in Python?
a) seed()
b) sqrt()
c) factorial()
d) print()

Q.2. What is the output of the expression round(3.145663)?
a) 3
b) 4
c) 5
d) 2

Q.3. The function pow(x, y, z) is evaluated as _____
a) (x**y)**z
b) (x**y)/z
c) (x**y)%z
d) (x**y)*z

Q.4. What is the output of the function complex(2+2j)?
a) Error
b) 1
c) 2
d) 2+2j

Q.5. Which of the function will not result in an error when no argument is passed to it?
a) min()
b) divmod()
c) all()
d) float()

Q.6. Python supports the creation of anonymous function at runtime using a construct called _____
a) lambda
b) pi
c) anonymous
d) None of the above

Q.7. What is the type of each element in sys.argv?
a) set
b) list
c) tuple
d) string

Q.8. What is return by math.ceil(3.4)?
a) 3
b) 4
c) 4.0
d) 3.0

Q.9. What is the order of namespaces in which Python looks for an identifier?
a) Python first searches the global namespace, then the local namespace, and finally the built-in namespace.
b) Python first searches the local namespace, then the global namespace, and finally the built-in namespace.
c) Python first searches the built-in namespace, then the global namespace, and finally the local namespace.
d) Python first searches the built-in namespace, then the local namespace, and finally the global namespace.

Q.10. Which of the following is false about the "import modulename" form of import?
a) The namespace of the imported module becomes part of the importing module.
b) This form of import prevents name clash.
c) The namespace of the imported module becomes available to the importing module.
d) The identifiers in the module are accessed as modulename.identifier.

Answers

Q.1. d) print()
Q.2. a) 3
Q.3. c) (x**y)%z
Q.4. d) 2+2j
Q.5. d) float()
Q.6. a) lambda
Q.7. d) string
Q.8. b) 4

Q.9.   c)   Python first searches the built-in namespace, then the global namespace, and finally the local namespace

Q.10.   a)   The namespace of the imported module becomes part of the importing module

Fill in the blanks

1.   These files are mostly written in _____format.
2.   A cursor is like a file handle we can use to perform_____ on the data stored in the database.
3.   Regular expression is used to_____ any substring based on some specific condition.
4.   SQLAlchemy is a/an _____tool to map database tables with Python objects, created by _____.
5.   JSON is a string with _____pair.

Answers

1.   ANSI textual(UTF-8)
2.   operations
3.   search
4.   ORM (object relational mapping), Mike Bayer
5.   key and value

6 Getting started with Kivy

Now, it's my belief that Python is a lot easier than to teach to students programming and teach them C or C++ or Java at the same time because all the details of the languages are so much harder. Other scripting languages really don't work very well there either.   ~Guido van Rossum

6.1 Why cross-platform?

Mobile applications are basically of two types – native apps and cross-platform. Native apps are developed only for specific platforms such as Android or iOS. But we cannot use the Android app in iOS or vice versa; therefore, cross-platform is the only solution to resolve it. Most of the cross-platform apps are being developed using HTML, CSS, and JavaScript. These are based on the interpreted programming language that causes poor performance and sometimes compatibility issues with the device.

6.2 What is Kivy

Kivy is a free and open-source Python library for developing mobile apps and other multitouch application software with a natural user interface (NUI). It is developed by Kivy organization along with Python for Android and Kivy OS. Kivy and BeeWare are the most popular frameworks used to design GUI applications. BeeWare is used to develop native apps for different platforms, but native apps have specific limitation, that is why we will focus on cross-platform app development. Kivy is a cross-platform framework to develop desktop and mobile applications. It was introduced in the year 2012; the latest version Python supports "write once and read many". Kivy uses the OpenGL library feature that makes it cross-platform. We can rapidly develop multitouch applications.

6.3 Why Python in mobile app

There are many choices available to develop a mobile app; Python being the easiest and most popular language is the preferred choice in mobile app development. Kivy is a beautiful toolkit to start GUI programming. It's based on Python, and Python is a great first language to learn programming even in the mobile app development. Moreover, we do not have to record our app for different platforms since Kivy supports Android, iOS, Desktop, Windows, Linux, and Mac. Also, we can use the same codebase for all platforms. Python comes with large libraries, packages, and frameworks to make the development easy. Python is one of the highest paid programming languages. It is a flexible and dynamic language with scalability and wide range of

https://doi.org/10.1515/9783110689488-006

tools for development. Due to portability, apps can work on various operating systems (Windows, Linux/Unix, iOS, Mac, Android, OS/2, Amiga, Raspberry Pi, etc.) and architecture. There is a wide availability of code over the Internet free of cost so the developer can use several cross-platform techniques to create a quality app. Some tools such as Whoop, Widgetpad, MoSync, MobiCart, RhoMobile, PhoneGap, and Cordova are also available to develop a mobile app. Big competitors of Kivy are Qt and Flash – these support multitouch too. These features make the competition very tough, but still, there are some big advantages of Kivy that makes Kivy the first choice to develop cross-platform apps.

Free – The major advantage is that it is free, being an open source there is no need to pay anything for redistribution or developing applications, it has code patch feature so that we can directly contribute in it.

Easy – We can develop software by writing a few lines of code, but others are not that easy. Kivy is exclusively written in Python, which makes it very convenient to use. Kivy has its description language to describe UI. Thus, we can design applications according to our own way.

Development support – Kivy is developed by professionals in the community. Some of the developers develop it for a living, which means it is not a small community worldwide. There is very little chance to get support on it.

Flexibility – It can run on various platform devices (Android, iOS, Windows, Linux, Raspberry pi). Its fast-paced development can adopt any technology, including the third party as well, any time, any module, such as WM_TOUCH(Windows), multi_touch(OS X), HID kernel(Linux), TUIO for other touch devices.

Fast – In this competitive environment, application development and its execution, both are very fast in Kivy. Kivy implements time-critical functionality on the C level. It has powerful algorithms to minimize big operations. In this channel, we support CPU, GPU, TPU, and so on. This feature can process 3D graphics with increased performance.

Design from scratch – Kivy is designed from scratch level to make an interaction between computer and human easy. We could face compatibility issues with any computer that is having different design and architecture, but with Kivy the possibility is remote because Kivy helps us to export executable code for each and every platform for the production environment.

6.4 Kivy architecture

Kivy architecture plays a significant role in software engineering (Figure 6.1). Here, we will discuss how everything is working with each other. It consists of many blocks.

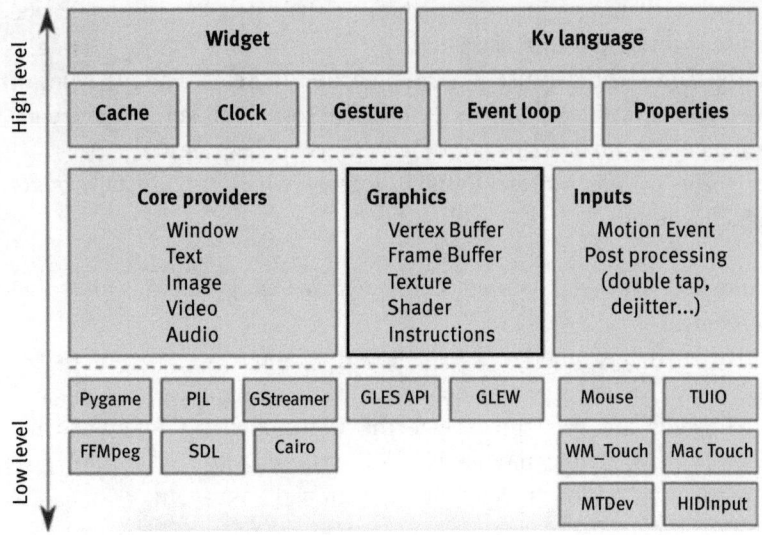

Figure 6.1: KivyaApp architecture.
Source: https://kivy.org/doc/stable/guide/architecture.html

Core providers and input providers – Basic or core tasks such as opening window, displaying images and text, playing audio, using the camera, spelling correction, handling a special pieces of code in all the platforms – this unique code is known as core provider. For example, Windows, Linux, and OSX are natively different platforms, but they can be handled by core providers. By using libraries that are shipped with any platform, Kivy efficiently reduces code in size for distribution and makes it easy to port to other platforms too. An interpreter is a piece of code that adds support for a specific input device, such as a mouse, keyboard, or touchpad, if we need to add support for the new input device, we can simply provide a new class that reads our input data from our device and transforms them into Kivy basic events.

Graphics – Kivy Graphics work on OpenGL at the lowest level. It issues hardware-accelerated drawing commands using OpenGL. Writing OpenGL code is a very difficult and confusing job. Kivy provides graphics API that can draw the things very quickly. Graphics API generates C level command for good performance and optimized code. The default version is OpenGL 2.0 ES (GLES2) on the entire device to fulfill all the cross-platform requirements.

Core – the provided features of core are as follows:
– **Clock –** We can use the clock to schedule timer event, which can be one-shot time and periodic timer.

- **Cache** – If we are willing to store something in the catch, we can use cache class instead of writing our own class.
- **Gesture detection** – We can use gesture detection to detect various types of strokes like circle or rectangle, or we can train it according to our requirement.
- **Kivy language** – We can describe our UI with Kivy language easily.
- **Properties** – This is different from Python property, which can link our widget code with the UI description.

UIX (Widget and Layouts) – it contains commonly used widgets and layouts that help to create UI quickly.

- **Widgets** – These are user-interface elements that we add to our program to provide functionality. They may or may not be visible, like Button, label, and list.
- **Layouts** – Layouts are used to arrange the UI components automatically. However, we can calculate components, positions, and size, but layouts provide them automatically, for example, Grid Layout and BoxLayout.

Modules – We can create small plugins, and these plugins can be used to inject into the Kivy program. We can inject our own modules whether they are added by Kivy developer or not.

Input events – Kivy supports input types such as touches, mic, TUIO, and so on. All these input devices support only 2D onscreen position with any individual input even. These inputs are handled by touch() class. A touch instance can be of any these states.

- **Down** – Touch is down only once, every moment where it first appears.
- **Move** – In this position touch can be for unlimited time. A move happens whenever the 2D position of touch changes.
- **UP** – We receive a touch event because nobody is going to hold a finger on the screen.

Widgets and event dispatching – Widget is an object that receives input events. All widgets are arranged in a tree. One widget can have any number of child widgets or https://kivy.org/doc/stable/_images/architecture.png none.

All the widgets are directly or indirectly children of widget class. When new input is available, Kivy sends it out in one event per touch. The root widget of the widget tree receives the event depending upon the state of the touched location. Each widget in the tree can choose either digest or pass the event on. The event handler returns valid means digested and handled correctly, or it passes the event to its child or parent class. As given in the code:

```
def on_touch_down(self, touch):
    if self.collide_point(*touch.pos):
        self.pressed = touch.pos
        return True
```

6.5 Python developed mobile apps

Dropbox – Dropbox is a cloud-based application where users can store their files, and the same can be mounted on the local machines too. As the user makes any changes in the local machine automatically, these changes reflect on the cloud too through the Internet.

YouTube – YouTube is one of the products from Google. YouTube is a platform where users can view and download videos along with generating logs such as likes, subscribers, and comments on a video.

Instagram – Instagram is a social media platform where users can share their views and see other user's views on a particular topic.

BitTorrent – It is a platform where users can share their files over the Internet, and anybody can download these files. Torrentis is a chain of files such as computer applications, videos, audios, and images.

Ubuntu Software Center – It is a platform like Google Play store, Apple store, and Microsoft store. Users using Ubuntu can download this useful application. Some of the software are paid by most of the applications that are available free of cost.

Aarlogic CO5/3 – GSM/GPS tracking PCB Python development board along with the support of Test Server based on Google Map.

App Backup – JailBroken iOS devices, uses to backup and restore settings and data of an App Store app.

Food Plus – Mobile food app to process food orders and tracking of orders.

PyRoute – A GPS capable mapping/routing application, gstreamer is an optional dependency that only needs to be installed if video display or audio is desired, ffpy-player is an alternative dependency for audio.

PyGame Library – Here we can build friendly mobile games easily similar to Kivy.

6.6 History of Kivy

Kivy is developed by Kivy organization, a very prestigious organization that has developed projects such as Python for Android, and Kivy iOS. Kivy comes with Cython because Cython is built-in with Kivy, and this is very important without that Kivy cannot run. Cython helps to generate highly optimized machine code for the device. Kivy is built over the Kivy language, also known as kv language. It is used to describe UI. The UI and event or actions can be created or added easily to it.

6.7 Why not Cordova

Cordova is an Apache project that is interwined with PhoneGap. We can say where PhoneGap ends, Cordova begins. Being newer, Cordova eventually produced a 1.8 Mc "Hello world.apk" a quarter size of the Kivy package. Though 20 times as big as a native Java app, deployed on Android, it initializes much faster than the Kivy app and comes up in portrait mode but rotates correctly if the users rotate their phone. As for the UI, Kivy should give us better performance as it works directly on OpenGL, but browser-based solutions would give more flexibility and probably easiest development.

6.8 Kivy versions

Kivy's supporting libraries are pygame, PIL, Cairo, and so on. Their installation is required according to the required version. The best way is to go with Cython and Pygame. Most of the problems can be reduced using this. Installation commands are given as follows:

```
$ pip install cython
$ pip install hg+http://bitbucket.org/pygame/pygame
$ pip install kivy
  Or
$ git clone https://github.com/kivy/kivy
$ cd kivy
$ make
```

6.9 How it is different from the cross-platform app framework

The cross-platform provides the ability to hit two targets with one arrow. Its feature is "Write code once and run on almost all the platforms," this sort of the frameworks earlier came with performance issues and erratic application behavior; they have now become mainstream as the cost of developing native apps for both the platforms is increasing day by day.

6.10 System requirement

All codes given in this book are valid for both python (excluding few examples) version but for the best result, tested environment is given in Table 6.1.

Table 6.1: System requirements.

Specification	Description
Python version	2.x(2.3.1 to 2.7.16) and 3.x(3.3.0 to 3.7.3)
Processor	Pentium-4 or higher
RAM	512 MB (1 GB recommended)
Operating system Screen resolution	Windows XP (including 64bit), Linux(Any), Mac OS X 10.5, Raspberry Pi, 800*600

6.11 Installation of Kivy

There are many ways to install Kivy on various platforms, some of them are mentioned later.

6.11.1 Installation in Windows?

Installation is straightforward in windows because of it being available in a bundle, and this bundle can be downloaded from the given URL http://kivy.org/#downlod. After completion of the download it will provide a zip file, that zip file can be extracted as shown in Figure 6.2. After this, we can find **kivy.bat** file, which will be used to execute our application. After executing this file, command line opens with Python interpreter with temporary environment settings.

6.11.2 Installation in Mac OS?

As shown in Figure 6.2, we can download **Kivy-1.1.1-osx.dmg** with all dependencies from the given website http://kivy.org/#download. After downloading, we can double click to install it and finally launch it from the launcher.

Operating System	File	Instructions	Size
Windows 7, 8, 10 (32/64 bit)	Python 2.7 and 3.5 to 3.7 is supported. Install using pip, follow the instructions here	Installation on Windows	--
OS X 10.9 or later	Install using pip, either using the system python (python2.7), or an installed python from 3.5 to 3.7. Or install using Kivy.app: Python 3 123.0MB OSX > 10.13.5 Python 2 92.7MB OSX > 10.13.5	Installation on macOS	
Linux (Ubuntu, Mageia, Arch, ...)	Python 3.5 to 3.7 wheels using pip, follow the instructions here. Or from source for Python 2.7, 3.5 to 3.7: Kivy-1.11.0.tar.gz (Mirror)	Installation on Linux	23 Mb
ANACONDA. Conda-Forge	Install using conda with conda-forge: `conda install kivy -c conda-forge`. Supports Windows, OSX, and Ubuntu. For audio/video support also install `gstreamer` and `gst-plugins-base` on OSX and Ubuntu, or `ffpyplayer` on all platforms.	--	--
Ubuntu PPA	Stable PPA Daily PPA	Installation on Ubuntu How to use software from PPA	12 Mb
OpenSUSE	--	Installation on OpenSUSE	
Fedora	--	Installation on Fedora	--
Android (>= 2.2, with OpenGL ES 2)	Kivy Laucher 1.9.0 (APK)	Packaging for Kivy Launcher	13 Mb
Raspberry Pi	KivyPie - Image for Raspberry Pi containing Kivy	Installation on Raspberry Pi	532 Mb
Slackware	SlackBuilds - Downloads for installing Kivy on Slackware	Installation on	--

Figure 6.2: Download web page from http//kivy.org/#download.
Source: http//kivy.org/#download

6.11.3 Installation in Linux?

According to Figure 6.3, we can download **Kivy-1.1.1.tar.gz** with all dependencies from the given website http://kivy.org/#download. After downloading we can set the environment variable for our operating system. Ubuntu users can use the following given command. For other versions, we can go to its website for all installation instructions:

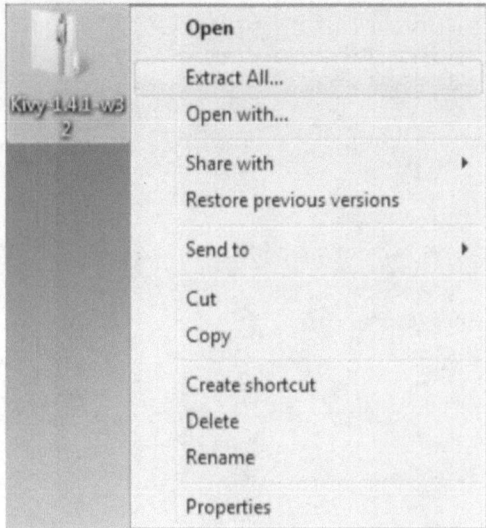

Figure 6.3: Kivy zip file extraction in windows.

```
$ sudo add-apt-repository ppa:kivy-team/kivy
$ sudo apt install python-kivy              //for python2
or
$ sudo apt install python3-kivy             //for python3
```

6.11.4 What are Wheels and pip?

These are unofficial windows binaries for Python extension packages. Wheels are pre-built distribution of a package that has already been compiled and does not require any additional step to install. We can download any file from https://www.lfd.uci.edu/~gohlke/pythonlibs/, it comes as **.whl** extension name. If we have whl file, it can be installed using the given command:

```
$ python -m pip install c:\Kivy-1.10.1-cp37-cp37m-win_amd64.whl
```

Kivy can be installed directly from the Internet as well. It will always install the latest version of Kivy. If we wish to install some specific version of Kivy, we can provide its version along with the command.

```
$ pip install kivy
or
$ pip install kivy-1.10.1
```

6.11.5 Cython installation in Linux?

According to the above-mentioned figure, we can download **Kivy-1.1.1.tar.gz** with

```
$ sudo add-apt-repository ppa:cython-dev/master-ppa
$ sudo apt update
$ sudo apt install cython
$ sudo apt install python-pip
$ sudo pip install -upgrade cython
```

6.11.6 Verifying installation

Installation can be verified using the given command:

```
>>>import pygame
pygame 1.9.6
Hello from the pygame community. https://www.pygame.org/contribution.html
```

6.12 Creating the First Hello World App

Creating an application in Kivy is very easy. We need to inherit App class into our application class, then we need to implement **build()** function/method, which always returns Widget instance (the root of our widget tree). Finally, we have to call **run()** function/method to the instance of our class. As shown in the given code (Figure 6.4):

```
>>>from kivy.app import App
>>>from kivy.uix.button import Button
>>>class MyFirstApp(App):
...     def build(self):
...             return Button(text= 'My First Cross Platform Application')
>>>MyFirstApp().run()
```

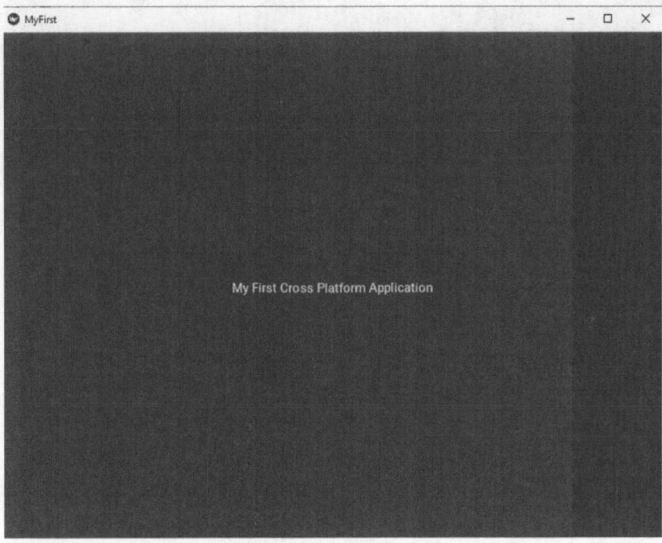

Figure 6.4: Hello World Program.

6.12.1 Event handling in Kivy

Events are an essential part of any programming language. Every event needs to be bound. This makes it easy to build whatever behavior we need. Figure 6.5 shows the given *event-handling architecture*.

6.12.2 Event dispatcher

It is the essential class of framework. We can register our event type and dispatch them to interested parties. **Widget, Animation, and clock** are event dispatchers. It depends on the main loop to generate and handle events.

6.12.3 Main loop

Loop is running in the entire lifetime of the application and stops when the application gets terminated. Every iteration of the loop event generated from user input, hardware sensors, and frames is rendered to display. If this callback called by our

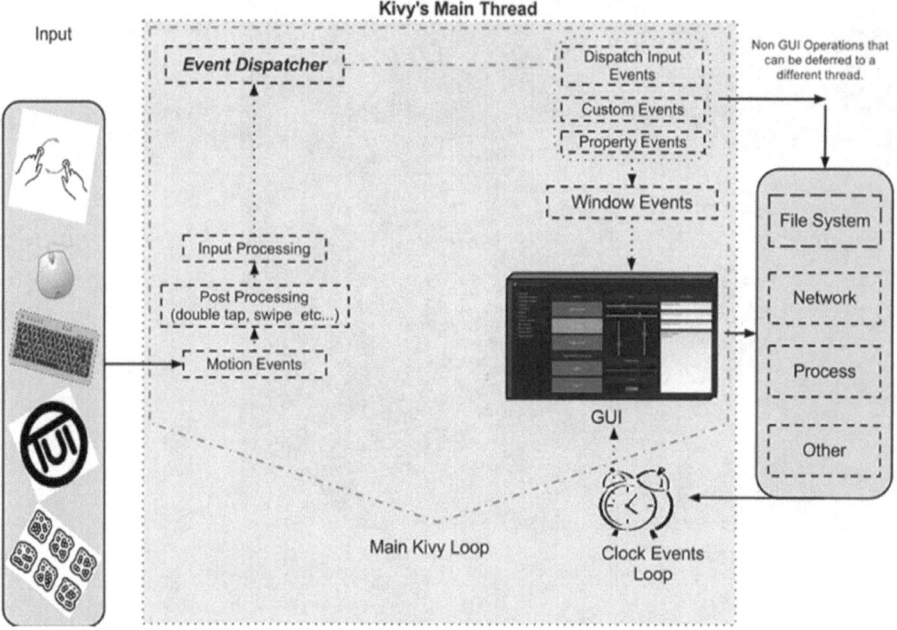

Figure 6.5: Event-handling architecture.
Source: https://kivy.org/doc/stable/_images/Events.png

application takes longer or does not arrive, it means our application will stop working. We must avoid long loops or sleeping. Let us see the examples:

```
>>>while True:
...    do_somthing()
...    time.sleep(0.5)
```

This loop will never stop until we stop our application.

6.12.4 Custom events

User can create their custom events by inheriting **EventDispacher** class, as shown in the given examples:

```
from kivy.event import EventDispatcher
class CustomEventDispatcher(EventDispatcher):
    def __init__(self, **kwargs):
        self.register_event_type('custom_event')
        super(CustomEventDispatcher, self).__init__(**kwargs)
```

```
def my_task(self, value):
    self.dispatch('custom_event', value)
def testEvent(self, *args):
    print ("An event has been dispatched", args)
def testCallbackEvent(value, *args):
        print ("An event has been dispatched", args)
ev = CustomEventDispatcher()
ev.bind(on_test=testCallbackEvent) #register event
ev.do_something('test')
```

6.13 Anatomy of Kivy

All the Kivy apps have one entry point **run()** method, as we have seen in the last code example. The essential class is App, which is available in kivy.app package that is why we wrote a line **from kivy.app import App** and this class became parent class of our application. In the next line, we wrote **from kivy.uix.button import import Button.** In uix package, we have all the UI components like layouts and widgets. **build()** is a default method to execute code and return an instance of the widget. Kivy has kv design language, which is like the HTML and CSS, where it is responsible for styling and adding elements to the display but does not handle any logic. There are specific rules that are defined for .kv file writing; some of them are given as follows:

- File must be saved with **.kv** extension name and the file name must be in lower case.
- The filename must match with the name of our main class and our main class ends with "*App"(lower or upper case).
- File must be saved in the same directory as our Python script.

```
---------------------------------my.kv---------------------------------
# Filename: my.kv
<MyGrid>:
    GridLayout:
        cols:2
        size: root.width, root.height
        GridLayout:
            cols:2
```

```
        Label:
            text: "Name: "
        TextInput:
            multinline:False
        Label:
            text: "Email: "
        TextInput:
            multiline:False
    Button:
        text:"Submit"
        on_press: app.btn()
```

```
import kivy
from kivy.app import App
from kivy.uix.label import Label
from kivy.uix.gridlayout import GridLayout
from kivy.uix.textinput import TextInput
from kivy.uix.button import Button
from kivy.uix.widget import Widget

class MyGrid(Widget):
    pass
class MyApp(App): # <- Main Class
    def build(self):
        return MyGrid()
if __name__ == "__main__":
    MyApp().run()
```

Figures 6.6 and 6.7 show the return Button instance.

6.13.1 Running Kivy application

We can write the displayed code in one file with an extension name **.py.** For example, our file name is **hello.py:**

```
#filename hello.py
from kivy.app import App
from kivy.uix.label import Label
```

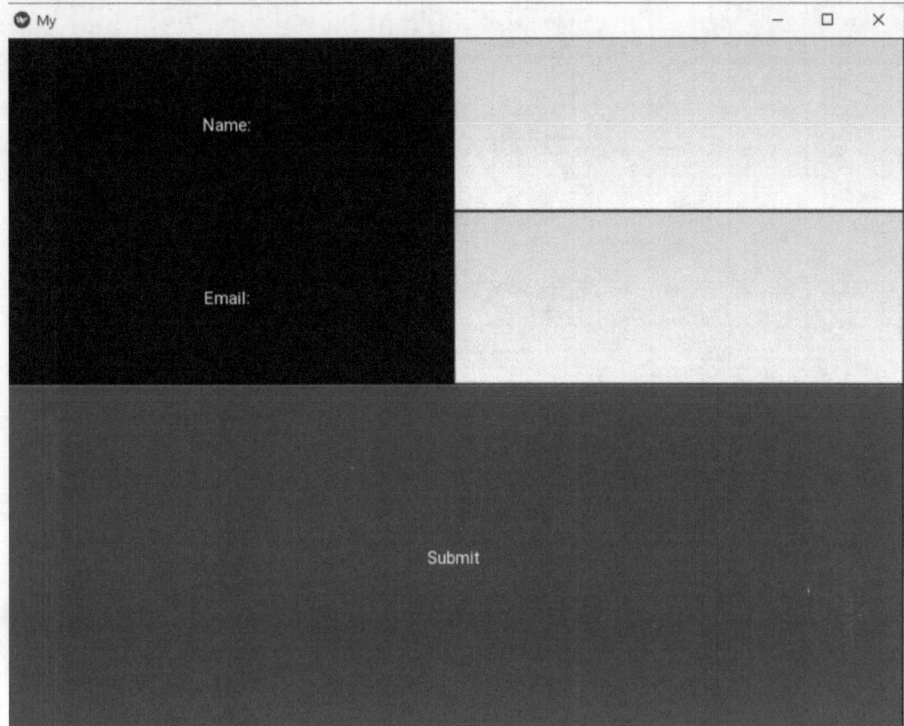

Figure 6.6: KIVY example.

```
class MyFirstApp(App):
    def build(self):
        return Label(text= "Testing")
MyFirstApp().run()
```

We can execute our application by using the following command:

```
$ python hello.py        #default python or python2
or
$ python3 hello.py        #for python3
```

To execute the same on Android, we require a different file. We can refer to topic packaging for Android. After execution, one window will open (Figure 6.8).

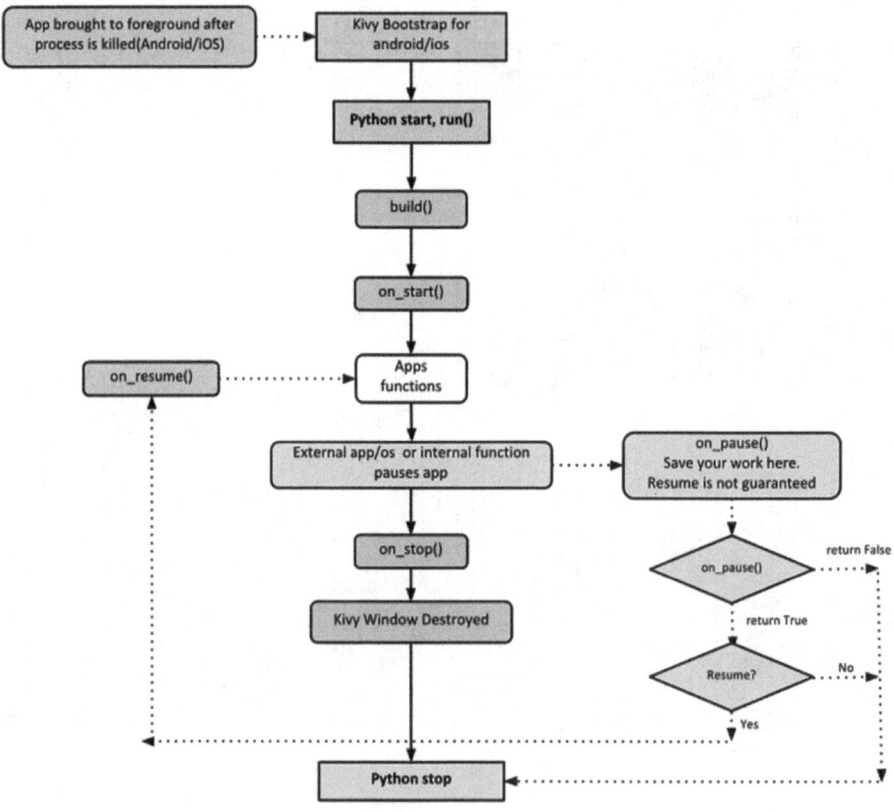

Figure 6.7: Life cycle of Kivy app.
https://kivy.org/doc/stable/_images/Kivy_App_Life_Cycle.png

6.13.2 Listing and uninstalling Kivy

Sometimes it becomes challenging to use the right version of Kivy because of multiple installations of Kivy versions. In this case, we need to list all the versions in the system, and then we can uninstall the earlier versions. The given command can be used to list all the installed Kivy versions into our system:

```
$ python -c 'import kivy; print(kivy.__path__)'
['/usr/local/lib/python3.6/dist/packages/Kivt-1.0.1-py3.7-linux-x86-64.
egg/Kivy', '/usr/local/lib/python3.6/dist/packages/Kivt-1.1.1-py3.7-linux
-x86-64.egg/Kivy', '/usr/local/lib/python3.6/dist/packages/Kivt-1.1.2-py
3.7-linux-x86-64.egg/Kivy']
```

```
$ sudo apt remove -purge python-Kivy
Or
$ sudo apt remove -purge python-Kivy-1.0.1
```

Figure 6.8: KIVY application.

Summary

- Mobile applications are of two types: native and cross-platform. Native applications are developed only for specific platforms such as Android and iOS, while cross-platform applications can play across all the operating systems.
- Kivy is an open-source and cross-platform library, which is developed for Python. It is developed by Kivy organization with the help of project Python for Android and kivy OS.
- We can design all the compatible device application using Kivy. Kivy applications are fast in performance (because of Cython). Kivy supports markup language that makes application development secure and easy.

- Kivy architecture is good to understand its internal mechanism. Examples of some most popular applications with their description are developed in Kivy such as Pygame, Pytorch, and so on.
- Kivy comes with Cython, which supports **kv** language. Kivy is better than Cordova. Various versions of Kivy are there with their installation system requirements.
- Configuring and running Kivy application on various platforms such as Windows, Mac, and Linux. Cython installation is done to generate optimized executable code and enables us to generate all machine-compatible code.
- Event handling in Kivy components, if anything happens like click, double click, and tapping. We can create EventDispatcher and custom event for components.
- Creating one sample application with its life cycle. The process to run a Kivy application from the command line. Listing and uninstalling Kivy from the system.

Key terms

Kivy architecture is essential to understand to develop any Kivy application. Cython plays a very important role in generating executable codes. Generally, thing code is not highly optimized by Cython, as it was developed to generate C language compatible code specific to any machine. Kivy installation and running application in Python with custom event handling. The custom events can be created to handle events by the programmer to avoid any mishappening during execution of the code.

Review questions

1. What is Kivy and its architecture?
2. How to create a custom event handler?
3. What are benefits of using Kivy over Cordova like cross-platform frameworks?
4. What is the main Kivy developed application?
5. What is Cython, and how it is suitable for the program?
6. How to install Cython?
7. What is the life cycle of Kivy application?

Exercise

Tick the correct option

Q.1. The cross-platform application _____
   a) works on Android
   b) works on iOS
   c) works on Windows
   d) works on all operating system

Q.2. Which is not a framework for mobile application development?
   a) Kivy
   b) Beeware
   c) Cordova
   d) SQLAlchemy

Q.3. Which of these makes Kivy apps platform independent?
   a) GNU
   b) OpenGL
   c) ASCII
   d) ANSI

Q.4. Which is not touch library in Kivy?
   a) WM_Touch
   b) Multi_Touch
   c) HID
   d) None of the above

Q.5. Which command is used to uninstall Kivy framework?
   a) purge
   b) delete
   c) kill
   d) force close

Q.6. Which is not the command to execute Kivy program?
   a) python abc.py
   b) python3 abc.py
   c) python2 abc.py
   d) All of the above

Q.7. What is the main entry point in Kivy app?
   a) run()
   b) main()
   c) MainString()
   d) None of the above

Q.8. What is the use of cache in Kivy?
   a) We can use cache class instead of writing our own class
   b) Its exception class
   c) Green computing
   d) Variable declaration

Q.9. What is the use of kivy.bat?
   a) To compile our application
   b) To execute our application
   c) To manage our application
   d) To delete our application

Q.10. What are "wheels" in Kivy?
   a) They are prebuild distribution of a package that has already been compiled
      and does not require any additional step to install
   b) They are car wheels
   c) They are is python libraries
   d) They are plugins

Answers

Q.1. d) works on all operating system
Q.2. d) SQLAlchemy
Q.3. b) OpenGL
Q.4. c) HID
Q.5. a) purge
Q.6. c) python2 abc.py
Q.7. a) run()
Q.8. c) Green computing
Q.9. b) To execute our application
Q.10. a) They are prebuild distribution of a package that has already been compiled
and does not require any additional step to install

Fill in the blanks

1. Kivy has a powerful support of _____ algorithms to minimize big operation. In this channel we support CPU, GPU, TPU, and so on.
2. Every iteration of the loop event generated from_____ is rendered to
   display.
3. Kivy helps us to export_____ for each platform for production
   environment.

4. We can describe our UI with _____ easily.

5. Inputs are handled by _____ class.

Answers

1. Cython C level
2. user input, hardware sensors, and frames
3. executable code
4. Kivy language
5. touch()

7 Kivy basics

> The only way to learn new programming language is by writing programs in it. ~Dennis Ritchie

7.1 Configuring environment

Kivy is an open source Python library for developing mobile apps and other multi-touch application software with a Natural User Interface (NUI). There are many more elements to the Kivi Language. We will learn how to organize elements with the help of Kivy API. We should be able to display most of the elements available for GUI design such as *Canvas* and *Page Layout*. Many environment variables are available to control the initialization and behavior of Kivy. It is recommended to install and use *home brew* (in Mac) or sudo pip (Ubuntu/other Linux), pip (Windows) to install Kivy on the *Terminal/Command Window in Linux, Mac and Windows*. The given command to restrict text render to the PIL (Pillow Library) implementation. This environment variable must be set before importing Kivy. If we are using PyCharm then most of the settings are done automatically on the default Python interpreter. PyCharm is available in two versions: Professional and Community. The Community version is free while the Professional version requires license fee to be paid:

```
$ sudo apt install python          #Command for Linux
$ KIVY_TEXT=pil python main.py      #Code execute are command
$ brew install python-kivy        #Kivy installation on python2
$ pip3 install --upgrade pip        #Upgrade pip version
$ pip install --upgrade pip       #Upgrade pip2
$ pip install python3-kivy        #Kivy installation on python3
$ pip install cython              #Cython installation to optimize code
------------------------------------------------------------------------
```

Each and every operating system has a different way to set the path, and it becomes necessary to set a variable using an environment variable. Therefore, we can go to set/get an environment variable value not only from the operating system but also from using a Python script as given in the following code:

```
import os
os.environ['KIVY_TEXT'] = 'pil'
os.environ['myname']='Tarkesh'
os.environ["KIVY_NO_ARGS"] = "1"
os.environ["KCFG_KIVY_LOG_LEVEL"] = "warning"
os.environ['KIVY_HOME'] = "/home/tarkeshwar/kivy"
```

https://doi.org/10.1515/9783110689488-007

```
print("KIVY_TEXT = ",os.environ['KIVY_TEXT'])
print("myname = ",os.environ['myname'])
print('KIVY_NO_ARGS = ',os.environ['KIVY_NO_ARGS'])
print('KCFG_KIVY_LOG_LEVEL = ',os.environ['KCFG_KIVY_LOG_LEVEL'])
print('KIVY_HOME = ',os.environ['KIVY_HOME'])

---------------------------------output---------------------------------
KIVY_TEXT = pil
myname = Tarkesh
KIVY_NO_ARGS = 1
KCFG_KIVY_LOG_LEVEL = warning
KIVY_HOME = /home/tarkeshwar/kivy
------------------------------------------------------------------------
```

The above-mentioned variables KIVY_TEXT, KIVY_NO_ARGS, KCFG_KIVY_LOG_
LEVEL and KIVY_HOME are environmental variables that are operated by the system.
These environment variables can be set in various ways in different operating systems. The best practice in production environment is by programming code. These
environmental variables problematically can be removed too with os.environ.pop
(**'myname'**). Following commands can be used to set environment variable:

```
$ touch ~/.bash_profile; open ~/.bash_profile # command to open bash shell
profile in linux
```

in the resulting window, we need to add new environment variable, as we can see
in the given file

```
$export MONGO_PATH=/Users/admin/Downloads/mongodb-osx-x86_64-4.0.6/bin
$export PATH=$PATH:$MONGO_PATH/bin
$export name=Tarkeshwar
$export PATH=$PATH:/Users/admin/Downloads/mongodb/bin./anaconda3/etc/pro-
file.d/conda.sh
$added by Anaconda3 2018.12 installer
$>>> conda init >>>
#!! Contents within this block are managed by 'conda init' !!
__conda_setup="$(CONDA_REPORT_ERRORS=false '/anaconda3/bin/conda' shell.
bash hook 2> /dev/null)"
if [ $? -eq 0 ]; then
\eval "$__conda_setup"
else
  if [ -f "/anaconda3/etc/profile.d/conda.sh" ]; then
    . "/anaconda3/etc/profile.d/conda.sh"
```

```
    CONDA_CHANGEPS1=false conda activate base
  else
    \export PATH="/anaconda3/bin:$PATH"
  fi
fi
unset __conda_setup
$<<< conda init <<<
```

Then we use command <u>source ~/.bash_profile</u> to apply all the settings without restarting or if we want to restart our PC it is up to us. The same can be implemented in Linux as well by using the following command:

```
$ vi ~/.bashrc
```

After adding this environmental variable, in the resulting file we can return back with saving our file, by pressing:**wq**. After restarting our machine, things will appear or we can reload settings by:

```
source ~/.bashrc
```

In windows, the environment variable can be set as a temporary or permanent variable too. Temporary variable can be set by the following command at the command line:

```
c:\>set path= "c:/Program Files/java/jdk1.8.1/bin"
c:\>set path= "c:/Program Files/kivy/bin"
```

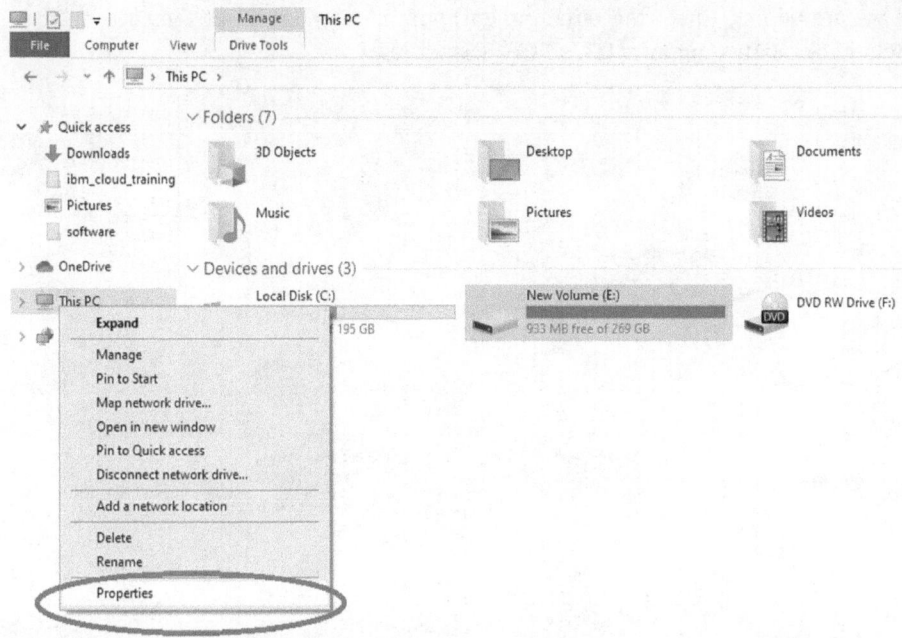

Figure 7.1: Setting environment variable in PC.

If we are willing to set the environment variable as permanent then we need to right-click on My Computer (This PC)→ Properties→

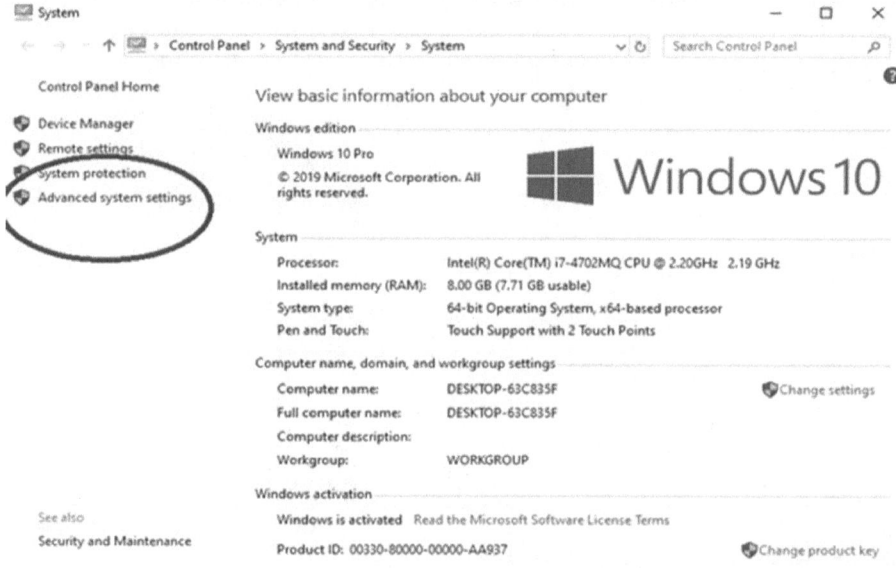

Figure 7.2: System property of PC.

Advanced system settings →

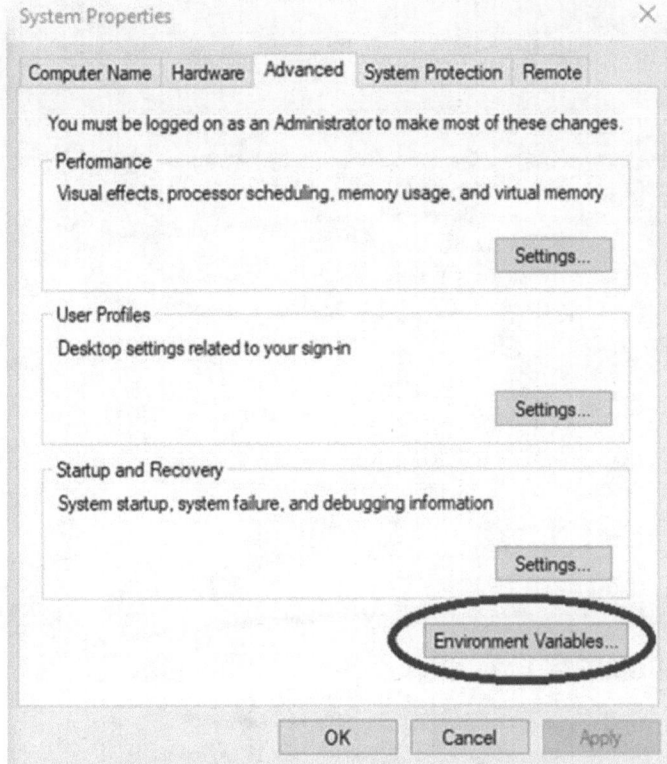

Figure 7.3: Setting environment variable in PC.

Environment Variable →

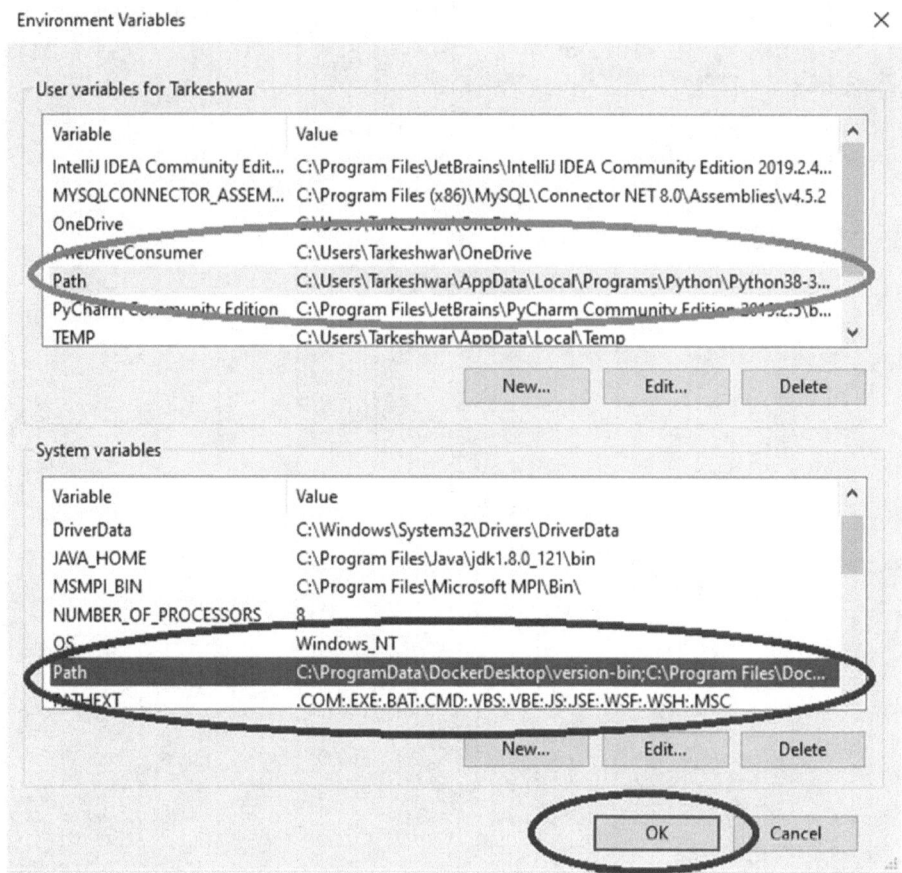

Figure 7.4: System environment variable in PC.

System Variable(System Wide) / User Variable (For Logged in User Only) →

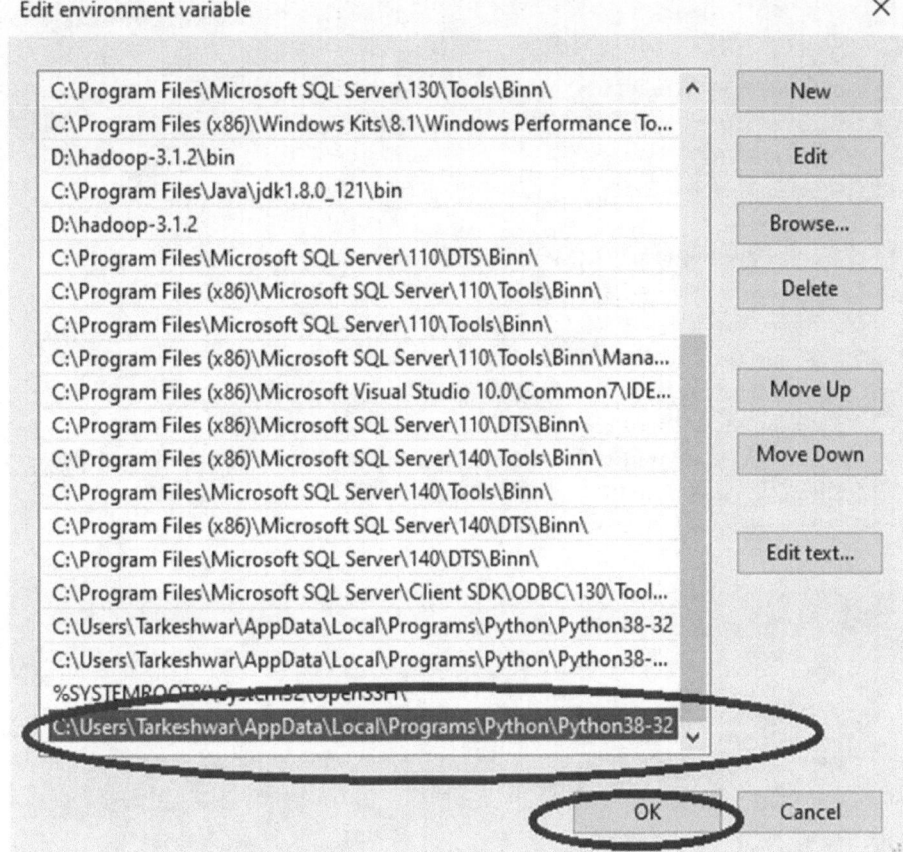

Figure 7.5: Adding program path.

Path. Finally, clicking on edit to make changes in the existing one, we can create new variables. The machine needs to be rebooted to use these environment variables to be replaced in the system. environment

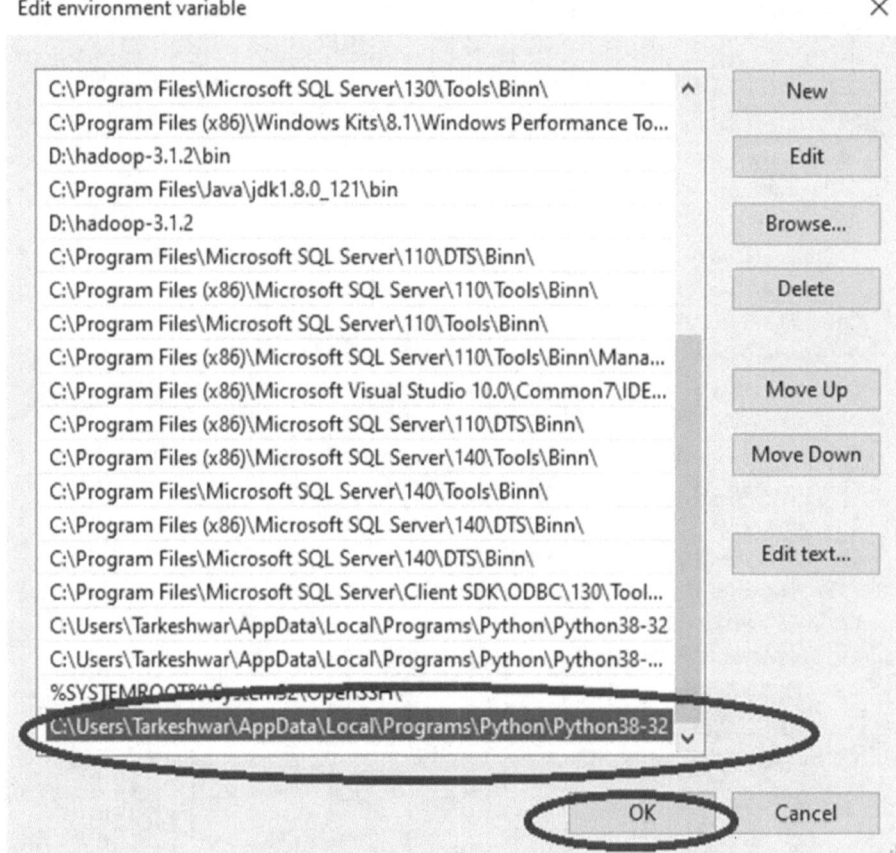

Figure 7.6: List of environment variable in PC.

Let us see the example sample program that uses scroll view. The scroll view is one of the commonly used layouts to scroll options; we will be creating one **scroll view** to navigate option easily. Displayed options are from the associated one excel data.xlsx file:

```
import kivy
kivy.require('1.0.8')
import openpyxl
from kivy.app import App
from kivy.uix.button import Button
from kivy.uix.scrollview import ScrollView
from kivy.uix.gridlayout import GridLayout
from kivy.uix.boxlayout import BoxLayout
```

```python
from kivy.app import runTouchApp
class AllApp(App):
    layout = BoxLayout(orientation='horizontal')
    layout_questions = GridLayout(cols=1, size_hint=(None, None),
    width=400, height=5500)
    def get_questions(self, subject):
        print(subject)
        wb = openpyxl.load_workbook('data.xlsx')
        sheet = wb[subject]
        data = []
        for row in range(2, sheet.max_row + 1):
            question = sheet.cell(row=row, column=1).value
            data.append(question)
        return self.add_questions(data)
    def add_questions(self, data):
        self.layout_questions.clear_widgets()
        for question in data:
            question = Button(text=question, size=(480, 40), size_hint=(1,
None))
            self.layout_questions.add_widget(question)
    def build(self):
        layout_subjects = GridLayout(cols=1, size_hint=(None, None),
        width=400, height=1000)
        subjects = ['Logic Building', 'SQL Server', 'Advanced Excel',
'Programming in Java', 'HTML5, CSS, JavaScript, JQuery', 'Web Development
using Servlet and JSP', 'Android Development using Java', 'Hibernate, Spring
and JSF', 'Application Testing using JUnit ', 'Programming using C#', 'Web de-
velopment using .Net Framework', 'Cross Platform app for Microsoft
PlayStore', 'distributed application with .net framwork', 'Machine Learning
using python', 'Big Data using R and Python']
        subjects_d = {}
        for subject in subjects:
            subjects_d[subject] = subject
        print(subjects_d)
        for k, v in subjects_d.items():
            k = Button(text=v, size=(480, 40), size_hint=(1, None))
            k.bind(on_press=lambda k: self.get_questions(k.text))
            layout_subjects.add_widget(k)
        root = BoxLayout()
        scroll_subject = ScrollView(size_hint=(1, 1))
        scroll_question = ScrollView(size_hint=(1, 1))
        root.add_widget(self.layout)
```

```
        self.layout.add_widget(scroll_subject)
        scroll_subject.add_widget(layout_subjects)
        self.layout.add_widget(scroll_question)
        scroll_question.add_widget(self.layout_questions)
        # runTouchApp(scroll_subject)
        # runTouchApp(scroll_question)
        return root
if __name__ == '__main__':
    AllApp().run()
```

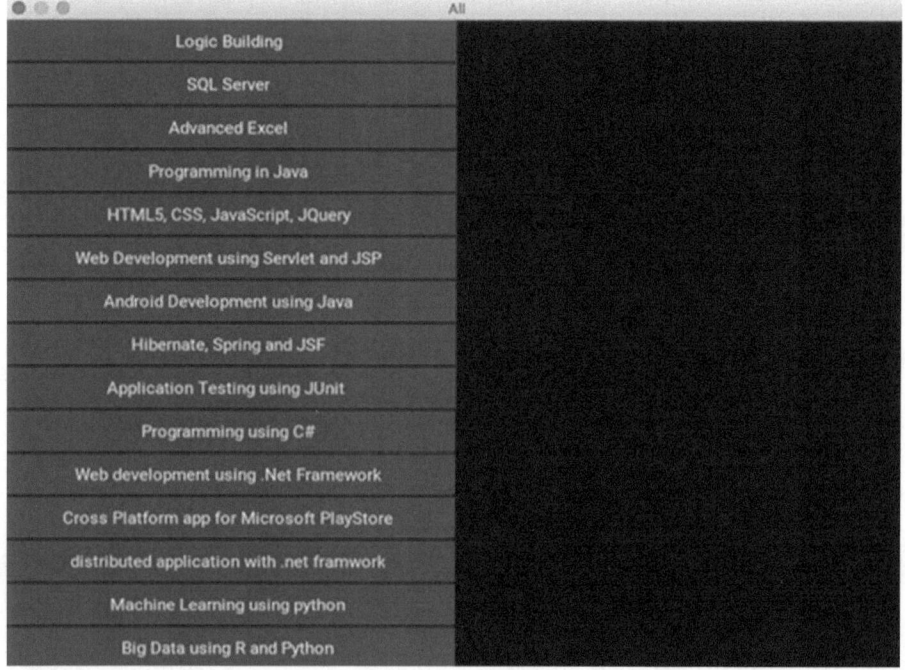

Figure 7.7: Scroll view display.

7.2 Configure Kivy and how to import Kivy library

Core tasks like opening window displaying images and text, playing audio and video, capturing images from the camera, spelling correction, make API both easy to use and extend. Configuration file for Kivy is named **config.ini**, and it adheres to the standard INI format. The location of the file is controlled by the environment variable KIVY_HOME. Sometimes it is desired to change the configuration only for certain applications or during testing of a separate part of Kivy such as input provider. We can create a separate configuration file using the given command as follows:

```
$ <HOME_DIRECTORY>/.kivy/config.ini
```

When we want to have a separate environment for garden, Kivy logs, and other things, we need to change the KIVY_HOME environment variable in our application to get the desired result. After doing this, the folder becomes like kivy local folder:

```
from kivy.config import Config
Config.read('config.ini')
# set config
Config.write()

--------------------------------config.ini----------------------------
[kivy]
keyboard_repeat_delay = 300
keyboard_repeat_rate = 30
log_dir = logs
log_enable = 1
log_level = info
log_name = kivy_%y-%m-%d_%_.txt
window_icon =
keyboard_mode =
keyboard_layout = qwerty
desktop = 1
exit_on_escape = 1
pause_on_minimize = 0
kivy_clock = default
default_font = ['Roboto', 'data/fonts/Roboto-Regular.ttf', 'data/fonts/
Roboto-Italic.ttf',  'data/fonts/Roboto-Bold.ttf',  'data/fonts/Roboto-
BoldItalic.ttf']
log_maxfiles = 100
window_shape = data/images/defaultshape.png
config_version = 21
[graphics]
display = -1
fullscreen = 0
height = 600
left = 0
maxfps = 60
multisamples = 2
```

```
position = auto
rotation = 0
show_cursor = 1
top = 0
width = 800
resizable = 1
borderless = 0
window_state = visible
minimum_width = 0
minimum_height = 0
min_state_time = .035
allow_screensaver = 1
shaped = 0
[input]
mouse = mouse
[postproc]
double_tap_distance = 20
double_tap_time = 250
ignore = []
jitter_distance = 0
jitter_ignore_devices = mouse,mactouch,
retain_distance = 50
retain_time = 0
triple_tap_distance = 20
triple_tap_time = 375
[widgets]
scroll_timeout = 250
scroll_distance = 20
scroll_friction = 1.
scroll_stoptime = 300
scroll_moves = 5
[modules]
[network]
useragent = curl
```

7.3 First app

Kivy depends on many libraries such as pygame, PIL, and gstreamer. As per system requirement, these libraries can be downloaded from their website or we can use pip command to install any library. Kivy application is very easy to create, we just have to keep in mind some of the rules such as sub-classing of the *App* class, implementing its *build()* function so it returns a widgets instance and instantiating the class and calling its *run()* function. Given code will pop up a black background with one button with text **"I can Display information over Screen."** A piece of code uses one of these specific APIs to talk to the operating system. The advantage of using specialized core provider for each platform is that we can fully leverage the functionality exposed by the operating system. This makes easier to port Kivy to another platform especially Android. It supports all the input devices such as Apple trackpad, mouse emulator, TUIO, and so on. The starting Python module for all Kivy applications should be named as **main.py** so that build tool can deploy automatically to mobile platform. As we see in the given example:

```python
import kivy
from kivy.app import App
from kivy.uix.button import Button
from kivy.uix.label import Label
kivy.require("1.0.6")

class LabelTestingApp(App):
    def build(self):
        return Button(text='I can Display information over Screen')

if __name__ == '__main__':
    LabelTestingApp().run();
```

Figure 7.8: Label demo.

In the above-mentioned example, we have used Python object orientation, but the same given application can be developed using kv language too. Kv language is a markup language like HTML. It is referred to as **kvlang**. We can design good-looking screens using it. UI files should be saved as application class name along with .kv extension name. In the above-mentioned example, application class name is **LabelTesting**, which should be written as **labeltesting** or **labelTesting**. So the file name will be like this **labeltesting.kv.** The Kivy file contains property in the form of key-value pair. If value is string, then it must be enclosed by double or single quotes.

```
import kivy
from kivy.app import App
from kivy.uix.button import Button
from kivy.uix.label import Label
kivy.require("1.0.6")
class LabelTestingApp(App):
    pass
if __name__ == '__main__':
    LabelTestingApp().run();
-----------------------------labeltesting.kv--------------------------
#filename labelTesting.kv or labeltesting.kv
Button:
    text:"I can Display information over Screen"
----------------------------------------------------------------------
```

One kv file can contain many components see in this code.

```
from kivy.app import App

class HelloApp(App):
    pass

if __name__=="__main__":
    HelloApp().run()

----------------------------------hello.kv----------------------------
#filename hello.kv
BoxLayout:
  orientation:"vertical"
  Button:
    text:'Button1'
  Button:
    text:'Button2'
  Button:
    text:'Button3'
```

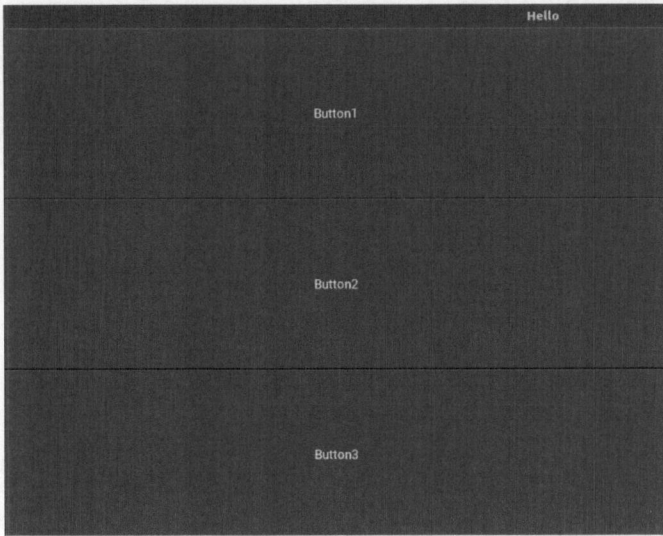

Figure 7.9: Multiple button demo.

By default, BoxLayout renders its child components horizontally(left to right), and orientation can be changed to vertical (top to bottom). KV parser is responsible to compile kv file. In case any error is found in it then it will raise **ParseException** that must be handled in production environment. The given exception is raised by traceback along with its line number in **kv** file.

7.3.1 Widgets

Widgets are used to describe any user interface (UI) element such as Label, Button, text input, file chooser, and tabbed boxes, and the layouts are used to layout these GUI components such as BoxLayout, GridLayout, and FloatLayout. If we are looking to create our own widget, just by extending that particular widget we can make our own implementation. "Any component can contain any other," can be achieved by using indentation in kv language. Outer most boxes are known as root widget; only one root widget is possible in one kv file. In the above-mentioned code, Button is the root widget.

Kivy widgets can be categorized as follows:

UX widgets – Classical UI widgets, ready to be assembled to create more complex widgets such as Label, Button, CheckBox, Image, Slider, Progress Bar, Text Input, Togge Button, Switch, and Video.

Layout – A layout widget does no rendering but just acts as a trigger that arranges its children in a specific way. Examples of Layouts are Anchor Layout, Box Layout, Float Layout, and Grid Layout.

Complex UX widgets – Nonatomic widgets are the result of combining multiple classic widgets. Its assembly and usages are not as generic as the classic widgets such as Bubble, Drop-down List, File Chooser, Popup, spinner, Recycler View, Tabbed Panel, Video Player, and Vkeyboard.

Behavior widgets – These widgets do not render but act on the graphics instructions or interaction(touch) behavior of the children such as Scatter, Stencil View, and so on.

Screen manager – Screen manager manages the screens and transitions when switching from one to another screen. Such as screen manager.

7.3.2 Custom widgets

We can create custom widget that encapsulates all of the required components as a single component. Inheritance provides capability by inheriting all the parent state and behavior. All the inherited state and behavior can be overridden according to requirement in the child class:

```
from kivy.app import App

from kivy.uix.widget import Widget
class MyCustomForm(Widget):
    pass
class MyApp(App):
    def build(self):
        return MyCustomForm()

if __name__=="__main__":
  MyApp().run()

----------------------------------my.kv---------------------------------
#filename my.kv

<MyCustomForm@BoxLayout>:
    orientation:"vertical"
    BoxLayout:
        orientation:"vertical"
        Button:
            text:'Add'
        Button:
```

```
        text:'Remove'
    FileChooserListView:
        id:filechooser
        #item_strings:["Tarkeshwar","Swapnil","Anand","Devendra"]
    BoxLayout:
        orientation:"vertical"
        TextInput:
                hint:'Address'
                multiline:False
```

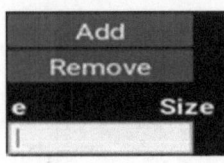

Figure 7.10: Custom widget.

MyCustomForm is a root widget. The @ symbol in Kv language indicates that class is extending BoxLayout using inheritance, which creates it as BoxLayout and angle bracket telling that it is a new class rule and not a root widget.

7.4 Orientation horizontal and vertical

Kivy creates an impressive amount of screen just using simple object-oriented code. BoxLayout arranges children in a vertical and horizontal box. The position widgets are above/below each other in vertical box. To position widgets next to each other we use a horizontal BoxLayout. In the given example, we have used 10 pixels spacing between children. The first button covers 70% of the horizontal space and second one 30% of available space. If orientation is vertical x, right and center_x will be used and if orientation is horizontal y, top and center_y. The size_hint uses the available space after subtracting all the fixed-size widgets. First button will be 200px wide and rest according to the available space. Let us see the sample code:

```
btn1 = Button(text='Hello', size=(200, 100), size_hint=(None,None))
btn2 = Button(text='Kivy', size_hint=(.5, 1))
btn3 = Button(text='World', size_hint=(.5, 1))

----------------------------Example---------------------------
import kivy.uix.boxlayout
import kivy.uix.textinput
import kivy.uix.label
import kivy.uix.button
```

```
from kivy.app import App

class SimpleApp(App):
    def build(self):
        self.textInput = kivy.uix.textinput.TextInput()
        self.label = kivy.uix.label.Label(text="Your Message.")
        self.button = kivy.uix.button.Button(text="Click Me.")
        self.button.bind(on_press=self.displayMessage)
        self.boxLayout = kivy.uix.boxlayout.BoxLayout(orientation="vertical")
        self.boxLayout.add_widget(self.textInput)
        self.boxLayout.add_widget(self.label)
        self.boxLayout.add_widget(self.button)
        return self.boxLayout

    def displayMessage(self, btn):
        self.label.text = self.textInput.text

if __name__ == "__main__":
    simpleApp = SimpleApp()
    simpleApp.run()
```

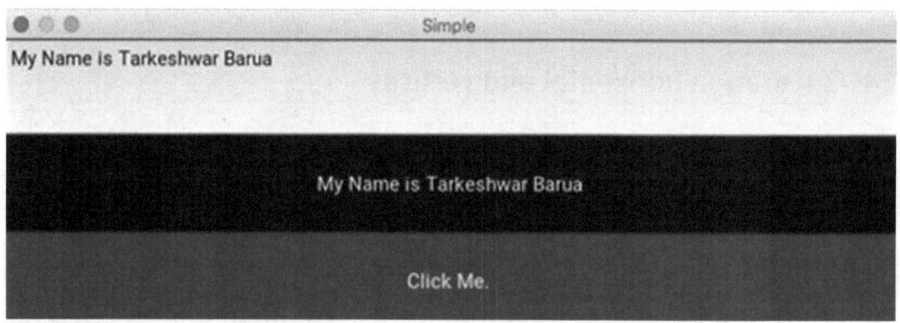

Figure 7.11: Orientation vertical.

Let us see another program that uses horizontal orientation for its component:

```
from kivy.app import App
from kivy.uix.boxlayout import BoxLayout
from kivy.uix.button import Button
class MainApplication(App):
    def build(self):
        layout = BoxLayout(orientation='horizontal')
        btn1 = Button(text='Hello')
```

```
        btn2 = Button(text='World')
        layout.add_widget(btn1)
        layout.add_widget(btn2)
        return layout

if __name__ =='__main__':
    MainApplication().run()
```

Figure 7.12: Orientation horizontal.

7.4.1 size-hint, pos_hint

There are plenty of problems one may face during rendering UI. The size_hint and position_hints are partially working, depending on the orientation. This is very easy to handle in kv language. If orientation is vertical vertical X and center_x will be used or Y top and center_y will be used. A horizontal BoxLayout will always make its children the same height as itself, no matter what the child widget requests. Child widget can make two types of size hint and absolute size, for each type of request, properties are X and Y dimension. All the given code in this book are valid for both Python versions but for the best result. A tested environment is given in the following example:

```
----------------------------------boxtest.kv----------------------------
# Filename: boxtest.kv
BoxLayout:
```

```
    orientation: 'vertical'
    Label:
        text: 'this on top'
    Label:
        text: 'this right aligned'
        size_hint_x: None
        size: self.texture_size
        pos_hint: {'right': 1}
    Label:
        text: 'this on bottom'
```

------------------------code with kv language--------------------------

```python
from kivy.app import App
class boxtestApp(App):
    pass

if __name__=='__main__':
    boxtestApp().run()
```

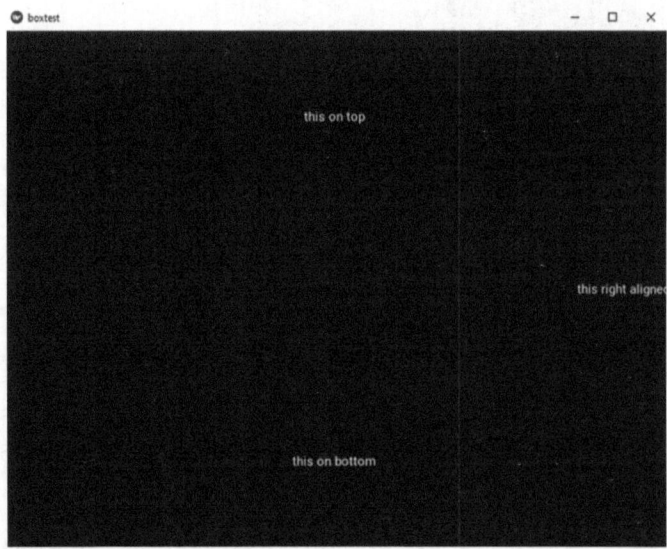

Figure 7.13: Component alignment.

The given orientation can be used to apply with horizontal and vertical. Let us have a look on the given code:

```
------------------------App with python code----------------------------
from kivy.app import App
from kivy.uix.boxlayout import BoxLayout
from kivy.uix.button import Button

class MainApplication(App):
    def build(self):
        layout = BoxLayout(orientation='vertical')
        btn1 = Button(text='Hello', size=(200, 100), size_hint=(None, None))
        btn2 = Button(text='Kivy', size_hint=(.5, 1))
        btn3 = Button(text='World', size_hint=(.5, 1))
        layout.add_widget(btn1)
        layout.add_widget(btn2)
        layout.add_widget(btn3)
        return layout

if __name__ =='__main__':
    MainApplication().run()
```

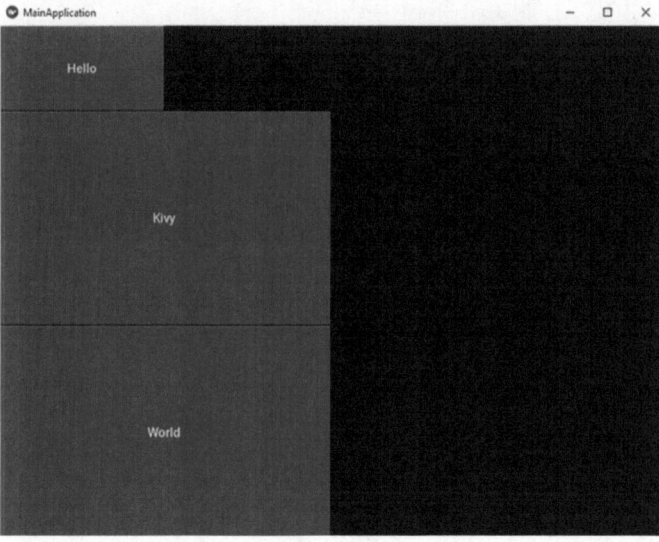

Figure 7.14: Vertical component alignment.

The size_hint is a proportional measure. If the three widgets have the same size_hint, they will be the same size. If one widget size_hint is twice as big as another widget, then it will be rendered at double the size. The size of widget is calculated based on the sum of the size_hint value for all the widgets. A single widget with a size_hint of 1 has no meaning unless we also know that its child widgets have a size_hint of 2 or 0.5. default value for size-hint is 1.0. We can disable it too just by setting None to it. For example, size-hint=(None, None). And other possible options are 1 cm, 0.75 in, and 50 dp to set the size. Going with relative rendering is always a good practice because various devices such as retina display and pixel can be three times.

7.4.2 pos_hint

post_hint plays a very important role to position any UI component, in absence of this, it is called dumb positioning. Components are rendering repeatedly because their current position is initializing point (0, 0), which is the bottom left corner and buttons default size is (100, 100). Let us see in given example:

```python
import kivy
from kivy.app import App
from kivy.uix.button import Button
from kivy.uix.widget import Widget
kivy.require("1.0.6")

class HelloApp(App):
    def build(self):
        root = Widget()
        b1 = Button(text='Button-1')
        b2 = Button(text='Button-2')
        root.add_widget(b1)
        root.add_widget(b2)
        return root

# HelloApp().run()
if __name__ == '__main__':
    HelloApp().run();
```

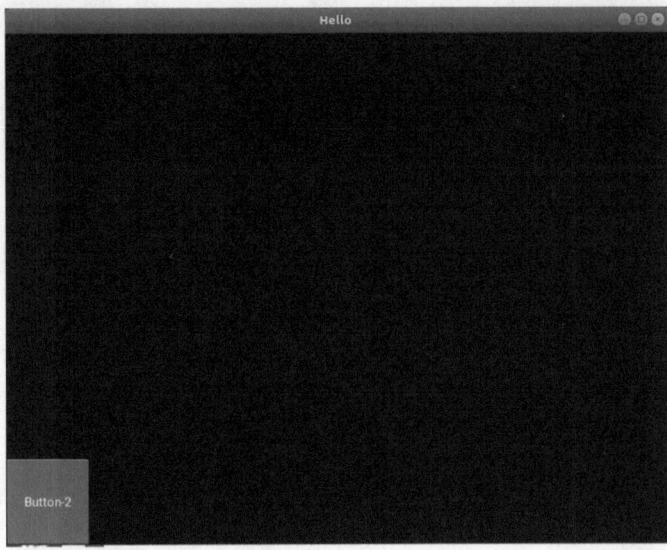

Figure 7.15: Button position 2D.

With the above-mentioned example, we study how to place a button, but some times in the industry it requires to place a button over and over. Then the following example will be useful:

```
import kivy
from kivy.app import App
from kivy.uix.button import Button
from kivy.uix.widget import Widget
kivy.require("1.0.6")

class HelloApp(App):
    def build(self):
        root = Widget()
        b1 = Button(text='Button-1', pos=(root.width,root.height))
        b2 = Button(text='Button-2',pos=(200,200))
        root.add_widget(b1)
        root.add_widget(b2)
        return root

# HelloApp().run()
if __name__ == '__main__':
    HelloApp().run();
```

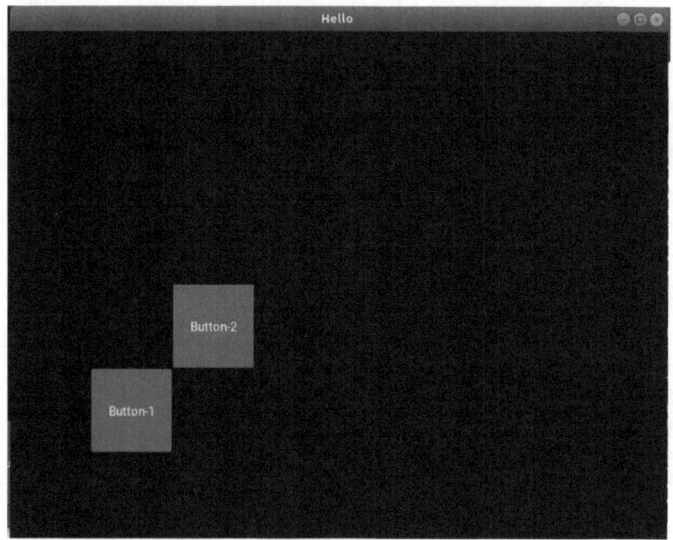

Figure 7.16: Button position overlapping.

7.4.3 FileChooser

If we want to choose files in the UI, Kivy provides widgets to do so using file chooser. We can display and browse files systems. It is one of the widgets, it supports Model view control design pattern. It can be customized according to our requirement. Let us see in the given code:

```python
from kivy.app import App
from kivy.uix.floatlayout import FloatLayout
from kivy.factory import Factory
from kivy.properties import ObjectProperty
from kivy.uix.popup import Popup
import os
class LoadDialog(FloatLayout):
    load = ObjectProperty(None)
    cancel = ObjectProperty(None)

class SaveDialog(FloatLayout):
    save = ObjectProperty(None)
    text_input = ObjectProperty(None)
    cancel = ObjectProperty(None)

class Root(FloatLayout):
    loadfile = ObjectProperty(None)
```

```python
        savefile = ObjectProperty(None)
        text_input = ObjectProperty(None)

    def dismiss_popup(self):
        self._popup.dismiss()

    def show_load(self):
        content = LoadDialog(load=self.load, cancel=self.dismiss_popup)
        self._popup = Popup(title="Load file", content=content,
        size_hint=(0.9, 0.9))
        self._popup.open()

    def show_save(self):
        content = SaveDialog(save=self.save, cancel=self.dismiss_popup)
        self._popup = Popup(title="Save file", content=content,
        size_hint=(0.9, 0.9))
        self._popup.open()

    def load(self, path, filename):
        with open(os.path.join(path, filename[0])) as stream:
            self.text_input.text = stream.read()
            self.dismiss_popup()

    def save(self, path, filename):
        with open(os.path.join(path, filename), 'w') as stream:
            stream.write(self.text_input.text)
            self.dismiss_popup()

class Editor(App):
    pass

Factory.register('Root', cls=Root)
Factory.register('LoadDialog', cls=LoadDialog)
Factory.register('SaveDialog', cls=SaveDialog)
if __name__ == '__main__':
    Editor().run()
```

```
---------------------------editor.kv---------------------------
#:kivy 1.1.0
#filename editor.kv
Root:
    text_input: text_input
    background:1,0,0,1
    BoxLayout:
        orientation: 'vertical'
        BoxLayout:
```

```
                    size_hint_y: None
                    height: 30
                    Button:
                        text: 'Load'
                        on_release: root.show_load()
                    Button:
                        text: 'Save'
                        on_release: root.show_save()
            BoxLayout:
                TextInput:
                    id: text_input
                    text: ''
                RstDocument:
                    text: text_input.text
                    show_errors: True
<LoadDialog>:
    BoxLayout:
        size: root.size
        pos: root.pos
        orientation: "vertical"
        FileChooserListView:
            id: filechooser
        BoxLayout:
            size_hint_y: None
            height: 30
            Button:
                text: "Cancel"
                on_release: root.cancel()
            Button:
                text: "Load"
                on_release:  root.load(filechooser.path,  filechooser.selec-
tion)
<SaveDialog>:
    text_input: text_input
    BoxLayout:
        size: root.size
        pos: root.pos
        orientation: "vertical"
        FileChooserListView:
            id: filechooserr
            on_selection: text_input.text = self.selection and self.selec-
tion[0] or ''
```

```
    TextInput:
        id: text_input
        size_hint_y: None
        height: 30
        multiline: False
    BoxLayout:
        size_hint_y: None
        height: 30
        Button:
            text: "Cancel"
            on_release: root.cancel()
        Button:
            text: "Save"
on_release: root.save(filechooser.path, text_input.text)
```

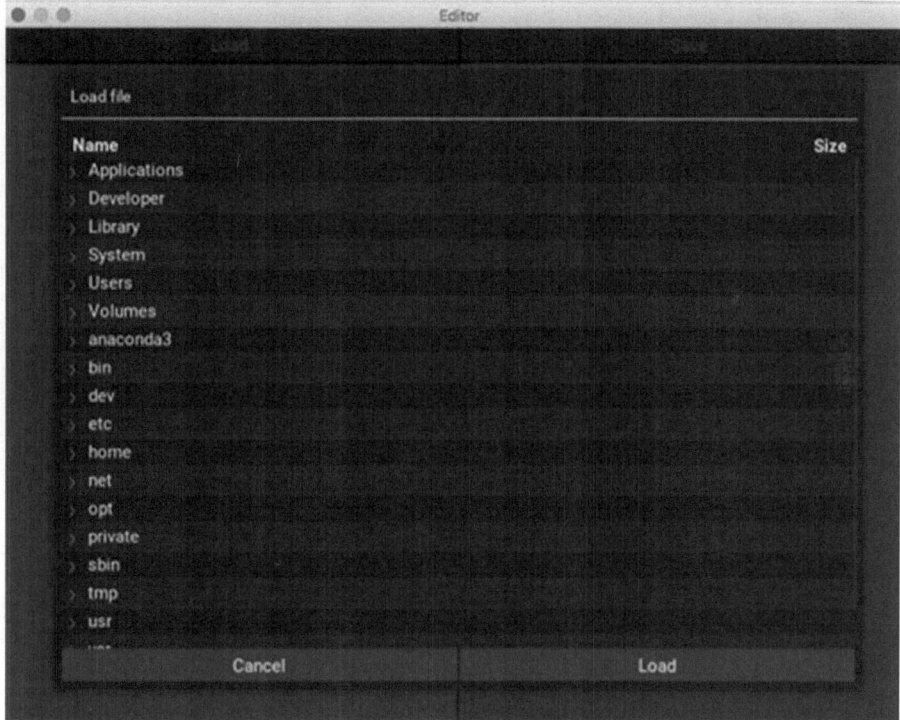

Figure 7.17: File chooser.

7.5 super, padding, __init__, add_widget

Padding is the space widget box and children; it can be defined in pixel such as padding_left, padding_right, padding_top, padding_bottom. To set all padding we can use only padding too, then the given padding will be applicable to all padding. It can accept two parameters too, which means horizontal and vertical.

7.5.1 Super

Super keyword is used to call any parent class state and behavior. Let us see in the given code where Vehicle is a parent class and Car, Bus, Bike are child classes. All the child classes are inheriting all parent class properties. However, these properties are being overridden and we call some properties (like constructor, functions and variables) from the child class using super keyword:

```python
class Vehicle(object):
    def __init__(self, model, max_speed):
        self.model = model
        self.max_speed = max_speed
        self.speed = 0

    def accelerate(self, speed_difference):
        self.speed += abs(speed_difference)
        self.speed = min(self.speed,self.max_speed)

    def slow_down(self, speed_difference):
        self.speed -= abs(speed_difference)
        self.speed = max(self.speed,-5)

    def __str__(self):
        return " "+self.model+" "+str(self.max_speed)

    def show_status(self):
        print("Vehicle is ",self.model, "Its speed is ",str(self.speed),-
" km/h")
class Car(Vehicle):
    pass
```

```python
class Bus(Vehicle):
    def slow_down(self, speed_difference):
        super().slow_down(speed_difference)
        self.speed=max(self.speed,0)

class Bike(Vehicle):
    def __init__(self, name,max_speed):
        max_speed=min((max_speed,30))
        super().__init__(name,max_speed)

def slow_down(self, speed_difference):
        super().slow_down(speed_difference)
        self.speed=max(self.speed,0)

    def show_status(self):
        print("Bike is ",self.model, "Its speed is ",str(self.speed)," km/h")

bmw=Car("BMW-X6",300)
print(bmw)
bmw.max_speed=400
bmw.model="BMW-X7"
print(bmw.speed)
bmw.accelerate(30)
print(bmw.speed)
bmw.slow_down(5)
print(bmw.speed)
--------------------------------output-----------------------------
BMW-X6 300
0
30
25
```

7.5.2 __init__

"init" stands for initialization. It is used to create user-defined constructor in the class. By default, all the classes have one default and zero parameterized constructor. This constructor is being created by the interpreter at runtime. To create our own constructor, init will help us. After creating customized constructor, interpreter will not create any constructor. In the first code we have given details about default constructor and in the second code customized constructor. dir function is used to list all the information about a object. Let us see in the given code:

```python
class Person:
    def __str__(self):
        return "Name : "+str(self.name)+" Address : "+str(self.address)+" Phone
        : "+ str(self.phone)
print (dir(Person))
```

```
--------------------------------output--------------------------
['__class__', '__delattr__', '__dict__', '__dir__', '__doc__', '__eq__',
'__format__', '__ge__', '__getattribute__', '__gt__', '__hash__',
'__init__', '__init_subclass__', '__le__', '__lt__', '__module__',
'__ne__', '__new__', '__reduce__', '__reduce_ex__', '__repr__', '__se-
tattr__', '__sizeof__', '__str__', '__subclasshook__', '__weakref__']
-------------------------------------------------------------------
```

Let us see one more example by calling class parameterized constructor and overriden function __str__ function:

```python
class Person:
    def __init__(self, name="No Name",address="No Address", phone =
    "00000000"):
        self.name = name
        self.address = address
        self.phone = phone

    def __str__(self):
        return "Name : "+str(self.name)+" Address : "+str(self.address)+"
        Phone : "+ str(self.phone)
person1=Person()
person2=Person("Tarkeshwar")
person3=Person("Tarkeshwar", "Monrovia")
person4=Person("Tarkeshwar", "Monrovia", "334423424")
print("Person1 ",person1)
print("Person2 ",person2)
print("Person3 ",person3)
print("Person4 ",person4)

print (dir(Person))
```

```
-------------------------------output-------------------------------
Person1 Name : No Name Address : No Address Phone : 00000000
Person2 Name : Tarkeshwar Address : No Address Phone : 00000000
```

Person3 Name : Tarkeshwar Address : Monrovia Phone : 00000000
Person4 Name : Tarkeshwar Address : Monrovia Phone : 334423424
['__class__', '__delattr__', '__dict__', '__dir__', '__doc__', '__eq__',
'__format__',   '__ge__',   '__getattribute__',   '__gt__',   '__hash__',
'__init__',   '__init_subclass__',   '__le__',   '__lt__',   '__module__',
'__ne__', '__new__', '__reduce__', '__reduce_ex__', '__repr__', '__se-
tattr__', '__sizeof__', '__str__', '__subclasshook__', '__weakref__']

7.5.3 add_widget

add_widget is used to all view (such as Button, Text, Radio, FileChooser etc.) into
the layout (such as BoxLayout, AnchorLayout, and GridLayout). Let us see the
example:

```python
import kivy.uix.boxlayout
import kivy.uix.textinput
import kivy.uix.label
import kivy.uix.button
from kivy.app import App
class SimpleApp(App):
    def build(self):
        self.textInput = kivy.uix.textinput.TextInput()
        self.label = kivy.uix.label.Label(text="Your Message.")
        self.button = kivy.uix.button.Button(text="Click Me.")
        self.button.bind(on_press=self.displayMessage)
        self.boxLayout = kivy.uix.boxlayout.BoxLayout(orientation="vert-
ical")
        self.boxLayout.add_widget(self.textInput)
        self.boxLayout.add_widget(self.label)
        self.boxLayout.add_widget(self.button)
        return self.boxLayout

    def displayMessage(self, btn):
        self.label.text = self.textInput.text

if __name__ == "__main__":
simpleApp = SimpleApp()
simpleApp.run()
```

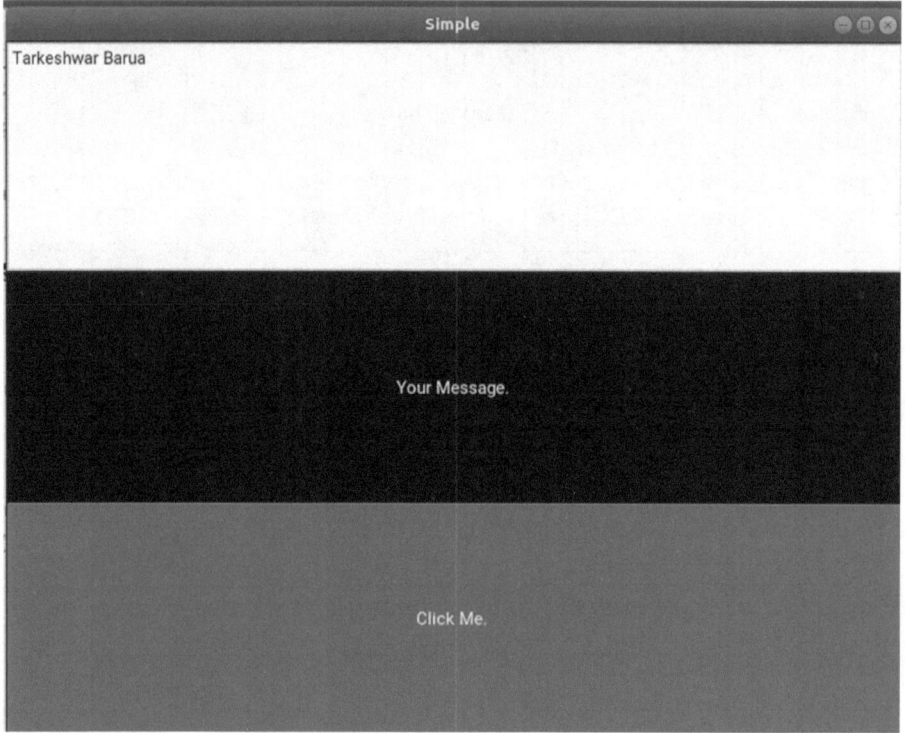

Figure 7.18: add_widget function.

Summary

- Configuring development environment is the first and basic step. There are various operating systems available, according to the operating system its installation is different too.
- Environmental variable plays a very important role during any configuration. We can use them into production environment on various platforms such as Windows, Linux, and Mac.
- Every Python written code file is known as module and Python supports write once and read many. This means we need to write our code only one time, and it can be used multiple times using import keyword.
- By default, it imports entire model code, but if we want a specific one than we can do by selecting one piece of code.
- Python supports Kv language too. Kv language is a markup language; using it we can create difficult UI within short period. We have created out first Kivy application using Python and Kv language.

- If mobile device user rotates its screen, then the mobile device needs to reload its UI again. This situation is known as orientation change. Kivy application supports both horizontal and vertical orientation. Default orientation is vertical.
- Position of the each and every component is fixed by machine because every device is different, to handle this condition we have size and pos hint. size_hint and pos_hint assures same type of screen rendered in all the devices.
- Super keyword is used to call properties from parent class.
- Padding is the space left between two or more than two components. Padding can be inherited from parent to child or directly to the child.
- Programmer can create their own customized widgets by inheriting, and properties can be modified by constructor too at the time of creating of object.
- add_widget id is used to add components to a layout instance. Any layout and any component can hold instances.

Key terms

Every Python code written code Python file is known as module and Python supports write once and read many, by default Python program imports entire model code but if we want specific code only then we can do by selecting one piece of code. Kv language is a markup language; using it we can create difficult UI within short period of time. If mobile device user rotates its screen, then the device needs to reload its UI again. This situation is known as orientation change. Default orientation is vertical. Position of the each and every component is fixed by machine because every device is different, to handle this condition we have size_hint and pos hint. These assure same type of screen rendered in all the devices. Super keyword is used to call properties from parent class. Padding is the space left between two or more than two components. Padding can be inherited from parent to child or directly to the child.

Review questions

1. What is the synonym of Python source file?
2. How to import module path from os package?
3. What is the major difference between Windows and Linux file system to make it cross-platform?
4. What is the escape sequence for new line and tab character?
5. What do DRY and SQL mean?
6. In which form is data stored inside relational database?
7. Constraints are usually applied to which part of a database?

Exercise

Tick the correct option

Q.1. What is the command to install Kivy _____?
a) brew
b) pip
c) sudo pip
d) All of the above

Q.2. Which of the following environment variables must be set before importing Kivy _____?
a) KIVY_HOME
b) JAVA_HOME
c) HOME
d) PYTHON_HOME

Q.3. Which of these versions of PyCharm is available free of cost _____?
a) PyCharm Professional
b) PyCharm Home
c) PyCharm Community
d) PyCharm Enterprise

Q.4. Which function of the os module is used to remove environment variable_____?
a) os.environ.pop('myname')
b) os.environ.popitem()
c) os.environ.remove('myname')
d) os.environ.clear('myname')

Q.5. Which file controls the configuration of our application _____?
a) Config.ini
b) requirements.txt
c) gitignore
d) LICENSE

Q.6. What is Kv language _____?
a) It is markup language
b) It is based on the HTML language
c) It is java language
d) It is Python language

Q.7. Which is not a widget _____?
a) Button
b) BoxLayout

c) FileChooser
d) Tabbed box

Q.8. By which OOP concept we can create own customized widget _____?
   a) Encapsulation
   b) Abstraction
   c) Polymorphism
   d) Inheritance

Q.9. How would you inherit in Kv language _____?
   a) CustomClass~ParentClass
   b) CustomClass$ParentClass
   c) CustomClass@ParentClass
   d) CustomClass^ParentClass

Q.10. What is the purpose of __init__ keyword _____?
   a) Creating function
   b) Creating variable
   c) Creating constructor
   d) Static constructor

Answers

Q.1. d) All of the above
Q.2. a) KIVY_HOME
Q.3. c) PyCharm Community
Q.4. a) os.environ.pop(**'myname'**)
Q.5. a) config.ini
Q.6. a) It is markup language
Q.7. b) BoxLayout
Q.8. d) Inheritance
Q.9. c) CustomClass@ParentClass
Q.10. c) Creating constructor

Fill in the blanks

1. Configuration file for Kivy is named _____.
2. In Kivy logs and other things, we will need to change the _____ environment variable in our application to get the desired result.
3. _____ supports all the input devices such as Apple trackpad, mouse emulator, and TUIO.

4. UI files should be saved as application class name along with _____ extension name.
5. A horizontal _____ will always make its children the same height as itself, no matter what child widget requests.

Answers

1. config.ini
2. KIVY_HOME
3. Kivy language
4. .kv
5. BoxLayout

8 Layouts

Python has been important part of Google since the beginning and remains so as the system grows and evolves. Today dozens of Google engineers use Python, and we are looking for more people with skill in this language.   ~Peter Norvig

8.1 What is the layout

Layouts are user rearrange user interface (UI) components, it is possible to calculate the component position and size. However, it is more convenient to use it from a machine because every machine is unique. Any layout widget does not render but it acts as a trigger that arranges its children in a specific way. Any layout can be nested. It is a special kind of widget that handles size and position of its child widgets. Various types of layouts have various ways to manage their child widgets. Almost all of them support the orientation of the screen like horizontal and vertical; when we change the orientation of mobile, they again draw over the screen. The same layout can be used in both orientations, but we can design different layouts too for another orientation to maintain better view rendered over the screen. size_hint indicates the size relative to the layout's size instead of absolute size. Values should be provided in the percentage-based tuple.

8.2 Box layout

Box layout arranges widgets in an adjacent way to fill all the available space. Its properties are size_hint, pos_hint. size_hint_y and size_hint_x it:

```python
from kivy.app import App
from kivy.properties import ListProperty
from kivy.uix.boxlayout import BoxLayout
from kivy.uix.button import Button
from kivy.uix.widget import Widget
class RootWidget(BoxLayout):
    def __init__(self, **kwargs):
        super(RootWidget, self).__init__(**kwargs)
        self.add_widget(Button(text='btn1'))
        cb = CustomBtn()
        cb.bind()
        self.add_widget(cb)
        self.add_widget(Button(text='btn2'))
    def btn_pressed(self, instance, pos):
```

https://doi.org/10.1515/9783110689488-008

```
        print ('pos: printed from root widget: {pos}'.format(pos=pos))
class CustomBtn(Widget):
    def _local_func(instance, pos):
        pressed = ListProperty([0, 0])
    def on_touch_down(self, touch):
        if self.collide_point(*touch.pos):
            self.pressed = touch.pos
            return True
        return super(CustomBtn, self).on_touch_down(touch)
    def on_pressed(self, instance, pos):
        print ('pressed at {pos}'.format(pos=pos))
class TestApp(App):
    def build(self):
        return RootWidget()
if __name__ == '__main__':
    TestApp().run()
```

Figure 8.1: Vertical orientation.

Let us see another example:

```
#filename LoadDialog.kv
BoxLayout:
    size:root.size
    pos:root.pos
    orientation:"vertical"
    FileChooserListView:
        size_hint:.6,.70
        pos:0,100
        id:filechooser
        BoxLayout:
            #size_hint_y:200
            height:10
            Label:
                color:1,0,0,1
                size_hint:.35,.10
                background_color:0,1,1,1
                font_size:50
                pos:0,0
                text:"Custom"
        Button:
        color:1,0,0,1
            size_hint:.35,.10
            background_color:0,1,1,1
            font_size:50
            pos:600,300
            text:"Cancel"
            on_release:root.cancel()
        Button:
            background_color:0,1,1,1
            text:"Load"
            font_size:50
            color:0,1,0,1
            size_hint:.35,.10
            pos:600,400
            on_release:root.load(filechooser.path, filechooser.selection)
            orientation:"vertical"
        TextInput:
            size_hint:.5,.1
            background_color:0,1,1,1
            font_size:40
```

```
                font_color:1,0,1,1
                pos:400,0

-------------------------------Main Program---------------------------
from kivy.app import App
from kivy.uix.floatlayout import FloatLayout
class LoadDialog(FloatLayout):
    def load(self,filename):
        print(filename)
    def cancel(self):
        pass
class LoadDialogApp(App):
    pass
LoadDialogApp().run()
```

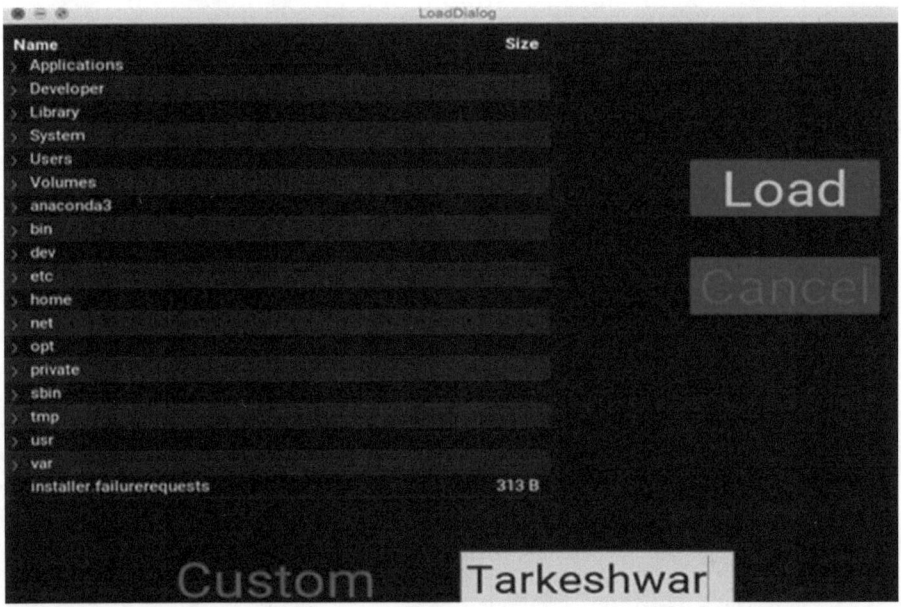

Figure 8.2: Box layout demo.

8.3 Float layout

It allows locating a child with arbitrary location and size, and it can be absolute or relative to the layout size. Default value is size_hint(1, 1) for every child. In case of more than one child, we can set **size_hint=None** and **pos_hint=None**, which means the machine will render components in absolute placing:

```
from kivy.app import App
from kivy.lang import Builder
root=Builder.load_string('''
FloatLayout:
    canvas.before:
        Color:
            rgba:0,1,0,1
        Rectangle:
            pos:self.pos
            size:self.size
    Button:
        text:'Click Me'
        text_size:self.size
        font_size:'50sp'
        markup:True
        size_hint:.5,.5
        pos_hint:{'center_x':.5,'center_y':.5}
''')

class MainApplicationApp(App):
    def build(self):
        return root
MainApplicationApp().run()
```

Figure 8.3: Float layout demo by internal KV.

Let us see another example:

```python
from kivy.app import App
from kivy.graphics.context_instructions import Color
from kivy.graphics.vertex_instructions import Rectangle
from kivy.uix.button import Button
from kivy.uix.floatlayout import FloatLayout
class RootWidget(FloatLayout):
    def __init__(self, **kwargs):
        super(RootWidget,self).__init__(**kwargs)
        self.add_widget(Button(text="Click Me", size_hint=(.5,.5),
            pos_hint={'center_x':.5, 'center_y':.5}))
class MainApp(App):
    def build(self):
        self.root=root=RootWidget()
        root.bind(size=self._update_rect, pos=self._update_rect)
        with root.canvas.before:
            Color(0,1,0,1)
            self.rect=Rectangle(size=root.size, pos=root.pos)
        return root

    def _update_rect(self,instance,value):
        self.rect.pos=instance.pos
        self.rect.size=instance.size
MainApp().run()
```

Figure 8.4: Float layout.

Let us see another code with external kv file:

```
#filename FloatTest.kv
FloatLayout:
    Button:
        text:"Button1"
        pos:100,100
        size_hint:.2,.4
    Button:
        text:"Button2"
        pos:200,200
        size_hint:.4,.2
    Button:
        text:"Button3"
        pos_hint:{'x':.3,'y':.6}
        size_hint:.5,.2
```

```
from kivy.app import App
class FloatTestApp(App):
    pass
FloatTestApp().run()
```

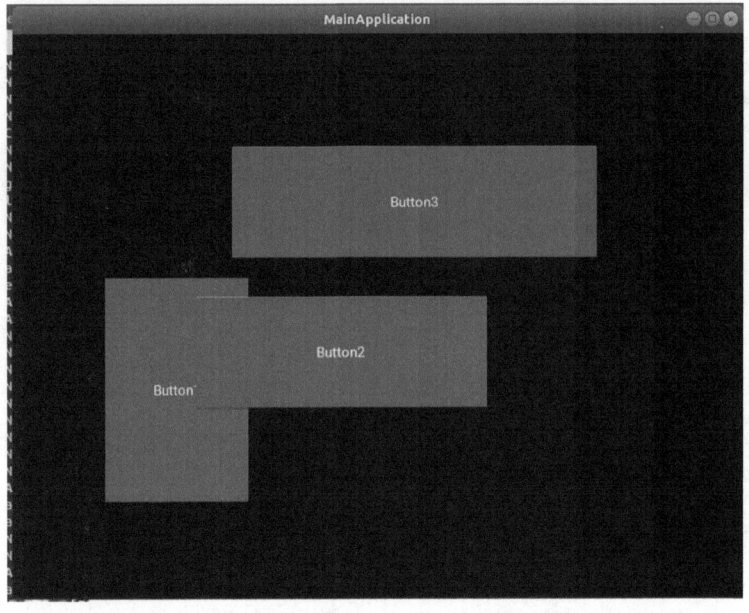

Figure 8.5: Float layout with buttons.

Let us see one more example:

```python
from kivy.app import App
from kivy.uix.button import Button
from kivy.uix.floatlayout import FloatLayout
from kivy.uix.gridlayout import GridLayout
from kivy.uix.image import Image, AsyncImage
from kivy.uix.label import Label
from kivy.uix.textinput import TextInput
class LoginScreen(FloatLayout):
    def on_touch_down(self, touch):
        for child in self.children[:]:
            if child.dispatch('on_touch_down', touch):
                return True
    def __init__(self, **kwargs):
        super(LoginScreen, self).__init__(**kwargs)
        self.cols=1
        self.add_widget(Image(source="dog.png"))
        self.add_widget(AsyncImage(source="dog.png"))
        self.add_widget(Label(text="User Name",font_size=40,
            pos_hint={'x':0.12,'y':0.15},
            pos=(200,200),
            size_hint=(0,1),
            color=(0.9,0.8,0,1)))
        self.username = TextInput(multiline=False,font_size=40,
            pos_hint={'x':0.3,'y':0.6},
            pos=(1,1),
            size_hint=(0.6,0.1))
        self.add_widget(self.username)
        self.add_widget(Label(text="Password",font_size=40,
            pos_hint={'x':0.11,'y':0.010},
            pos=(200,100),
            size_hint=(0,1),
            color=(0.9,0.8,0,1)))
        self.password = TextInput(password=True, multiline=False,font_size=40,
            pos_hint={'x':0.3,'y':0.45},
            pos=(1,1),
            size_hint=(0.6,0.1))
        self.add_widget(self.password)
        self.add_widget(Button(text="Login",font_size=40,
            pos_hint={'x':0.1,'y':0.2},
```

```
                pos=(200,100),
                size_hint=(0.4,0.2),
                color=(0.9,0.8,0,1),
                background_color=(0,1,1,1)))
        self.add_widget(Button(text="Cancel",font_size=40,
                pos_hint={'x':0.5,'y':0.2},
                pos=(200,100),
                size_hint=(0.4,0.2),
                color=(0.9,0.8,0,1),
                background_color=(0,1,1,1)))
class MyFirstAppApp(App):
    def build(self):
        return LoginScreen()
if __name__ == "__main__":
    MyFirstAppApp().run()
```

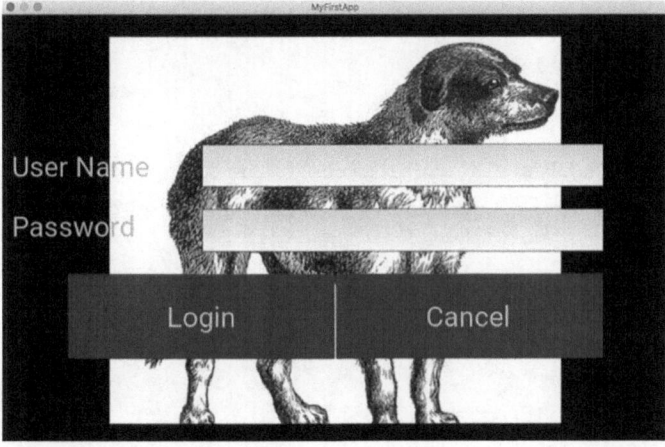

Figure 8.6: Float layout example.

8.4 Anchor layout

Anchor layout allows arranging the child at a position relative to the border of the layout. size_hint is not valid here. It only cares for its children position. This position is relative to a border(left, right, top, bottom, center) of the layout. By default, it is "center" for both X and Y where x is horizontal, and y is vertical. As we can see in this code:

```python
from kivy.app import App
from kivy.properties import ListProperty
from kivy.uix.anchorlayout import AnchorLayout
from kivy.uix.boxlayout import BoxLayout
from kivy.uix.button import Button
from kivy.uix.widget import Widget
class RootWidget(AnchorLayout):
    def __init__(self, **kwargs):
        super(RootWidget, self).__init__(**kwargs)
        cb = CustomBtn()
        cb.text="Hello"
        cb.bind()
        self.add_widget(cb)
        self.add_widget(Button(text='Button',
            font_size=40,
            pos_hint={'x':0.2,'y':0.2},
            pos=(200,100),
            size_hint=(0.5,0.2),
            color=(0.9,0.8,0,1),
            background_color=(0,1,1,1)))
    def btn_pressed(self, instance, pos):
        print ('pos: printed from root widget: {pos}'.format(pos=pos))
class CustomBtn(Widget):
    #super.height = 10
    #super. width = 10
    pos = 100,100
    text="Custom Button"
    def _local_func(instance, pos):
        instance.pressed = ListProperty([0, 0])
    def on_touch_down(self, touch):
        if self.collide_point(*touch.pos):
            self.pressed = touch.pos
            return True
        return super(CustomBtn, self).on_touch_down(touch)
    def on_pressed(self, instance, pos):
        print ('pressed at {pos}'.format(pos=pos))
class TestApp(App):
    def build(self):
        return RootWidget()
if __name__ == '__main__':
    TestApp().run()
```

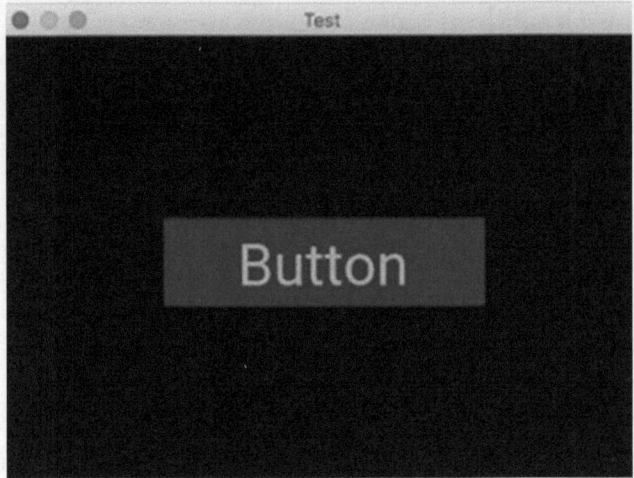

Figure 8.7: Anchor layout example.

Let us see another code:

```python
from kivy.app import App
from kivy.uix.anchorlayout import AnchorLayout
from kivy.uix.button import Button
class MyAnchorApp(App):
    def build(self):
        layout = AnchorLayout(anchor_x="left", anchor_y="center")
        btn = Button(text='Hello World',
            font_size=40,
            pos_hint={'x':0.2,'y':0.2},
            pos=(200,100),
            size_hint=(0.5,0.2),
            color=(0.9,0.8,0,1),
            background_color=(0,1,1,1))
        layout.add_widget(btn)
        return layout
if __name__=="__main__":
    MyAnchorApp().run()
```

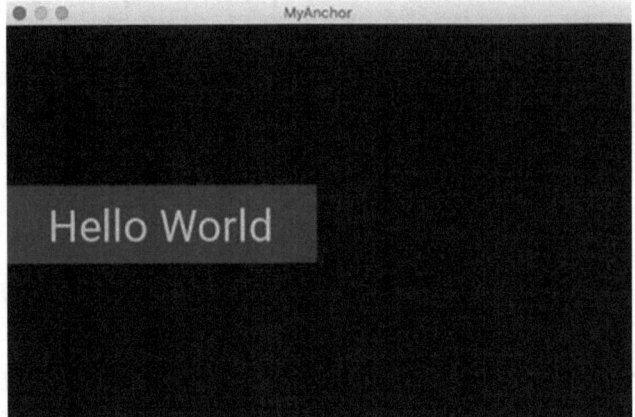

Figure 8.8: Anchor layout.

8.5 Grid layout

Grid layout is used to layout UI components in rectangular/tabular/grid format in specified rows and columns, a group that places its child screens in a rectangular grid that can be scrolled. We must specify at least one dimension of the grid so Kivy can compute the size of the element and decide how to arrange them. Let us see the given code:

```
from kivy.app import App
from kivy.uix.button import Button
from kivy.uix.gridlayout import GridLayout
from kivy.uix.image import Image
from kivy.uix.label import Label
from kivy.uix.textinput import TextInput
class LoginScreen(GridLayout):
    def __init__(self, **kwargs):
        super(LoginScreen, self).__init__(**kwargs)
        self.cols=1
        self.add_widget(Image(source="dog.png"))
        self.add_widget(Label(text="User Name"))
        self.username = TextInput(multiline=False)
        self.add_widget(self.username)
        self.add_widget(Label(text="password"))
```

```
        self.password = TextInput(password=True, multiline=False)
        self.add_widget(self.password)
        self.add_widget(Button(text="Login"))
        self.add_widget(Button(text="Cancel"))
class MyFirstAppApp(App):
    def build(self):
        return LoginScreen()
if __name__ == "__main__":
    MyFirstAppApp().run()
```

Here GridLayout class is used as a base class for root widget; we have overloaded the method __init__(), to add widget and add defined behavior. This thing is achieved by super method. The resulting screen is completely responsive, it means it will arrange UI components according to the available space. This is because of size hinting.

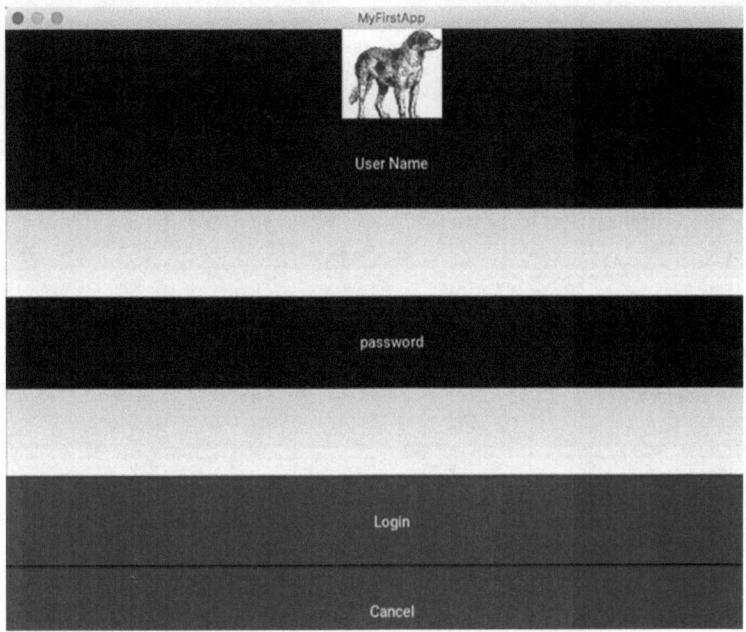

Figure 8.9: Grid layout.

We can see the same example by kv language too:

```python
import kivy
from kivy.app import App
from kivy.uix.label import Label
from kivy.uix.gridlayout import GridLayout
from kivy.uix.textinput import TextInput
from kivy.uix.button import Button
from kivy.uix.widget import Widget
class MyGrid(Widget):
    pass
class MyApp(App): # <- Main Class
    def build(self):
        return MyGrid()
if __name__ == "__main__":
    MyApp().run()
```

```
--------------------------------my.kv--------------------------------
# Filename: my.kv
<MyGrid>:
    GridLayout:
        cols:1
        size: root.width, root.height
        GridLayout:
            cols:2
            Label:
                text: "Name: "
            TextInput:
                multinline:False
            Label:
                text: "Email: "
            TextInput:
                multiline:False
            Button:
                text:"Submit"
                on_press: app.btn()
```

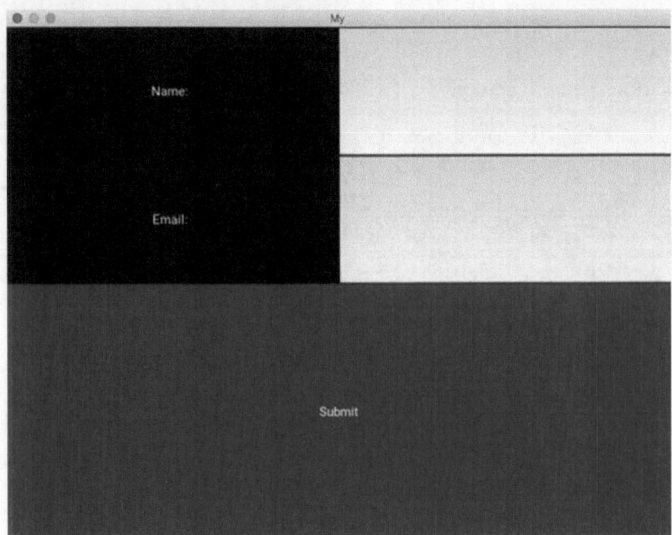

Figure 8.10: Grid layout demo.

8.6 Relative layout

Just like the Float layout, excluding positions are related to layout, not the screen. It allows to set relative coordinates for widgets. It supports the absolute positioning of the component. The child widget coordinates remain 0, 0 as they are always relative to the parent layout:

```
from kivy.app import App
class TwoApp(App):
    pass
TwoApp().run()

--------------------------------two.kv--------------------------------
BoxLayout:
    Label:
        text: 'Left'
    Button:
        text: 'Middle'
        on_touch_down: print('Middle: {}'.format(args[1].pos))
    RelativeLayout:
        on_touch_down: print('Relative: {}'.format(args[1].pos))
```

```
Button:
    text: 'Right'
    on_touch_down: print('Right: {}'.format(args[1].pos))
```

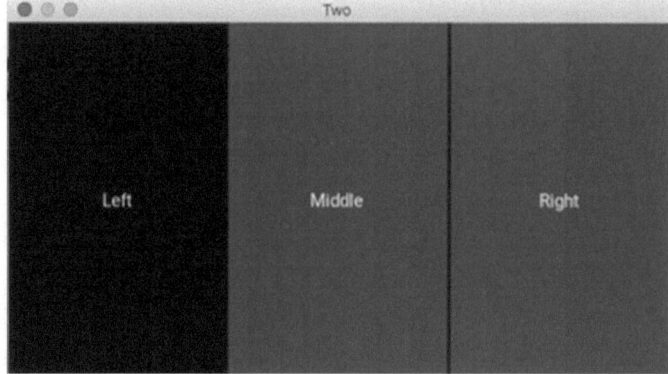

Figure 8.11: Relative layout.

```
Relative: (91.0, 158.99999999999997)
Right: (-287.6666666666667, 158.99999999999997)
Middle: (91.0, 158.99999999999997)
Relative: (305.99999999999994, 146.0)
Right: (-72.66666666666674, 146.0)
Middle: (305.99999999999994, 146.0)
Relative: (472.99999999999994, 193.0)
Right: (94.33333333333326, 193.0)
```

Figure 8.12: Output in the background.

8.7 Stack layout

It arranges child over each other. It can be used to design games based on image clicks scrolling bottom and top. This is useful to display child in a predefined size. We need to fix the size first. It arranges widgets to one another but with a set size in one of the dimensions, without trying to make them fit within the entire space. It is useful to display widgets of the same predefined size:

```
from kivy.app import App
from kivy.uix.button import Button
from kivy.uix.label import Label
```

```
from kivy.uix.pagelayout import PageLayout
from kivy.uix.stacklayout import StackLayout
from kivy.uix.widget import Widget
class MainApp(App):
    def printme(self,a):
        print("Yes Button has been Clicked",a)
        a.color = (1, 1, 0, 0)
    def printed(self,b):
        print("button released",b.text)
    def statetest(self,x,y):
        print("State Geting Changed")
    def colorchange(self,instance):
        print("instance ",instance)
        instance.color=(1,1,0,0)
    def build(self):
        btn=Button(text="Click Me",
            font_size=30,
            color=(1, 0, 0, 1),
            size_hint=(0.2,0.2),
            background_color=(0, 1, 0, 1))
        lbl = Label(text="Good Morning",
            font_size=40,
            color=(0, 1, 0, 1),
            #size_hint=(0.2, 0.2),
            #background_color=(0, 0, 1, 1))
        lbl.bind(on_press=self.colorchange)
        btn.bind(on_press=self.printme,
            on_release=self.printed,
            state=self.statetest)
        layout=StackLayout()
        layout.add_widget(btn)
        layout.add_widget(lbl)
        return layout
if __name__=='__main__':
    MainApp().run()
```

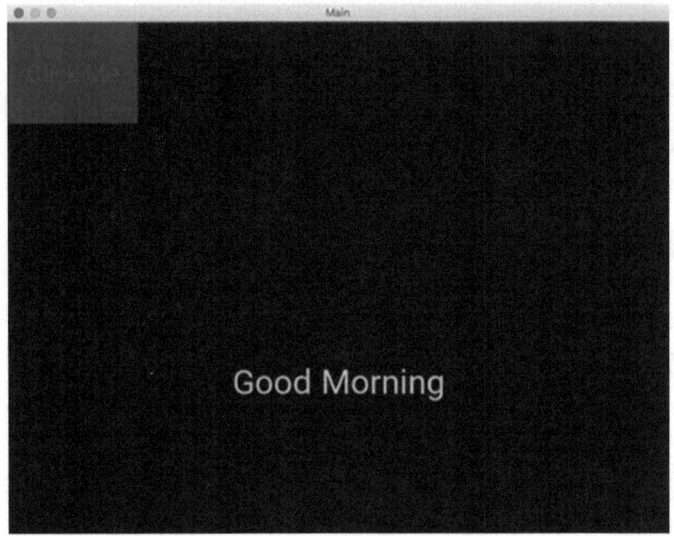

Figure 8.13: Stack layout.

8.8 Page layout

Page layout is entirely different to render layout dynamically. Components are stacked on each other, they flip through the pages, which means we can see the first added screen only. Multiple pages can be swaped with each other from left to right or right to left. Page layout does not support *size_hint, size_hint_min, size_-hint_max, pos_hint*:

```
from kivy.app import App
from kivy.uix.button import Button
from kivy.uix.label import Label
from kivy.uix.pagelayout import PageLayout
from kivy.uix.widget import Widget
class MainApp(App):
    def printme(self,a):
        print("Yes Button has been Clicked", a)
    def printed(self,b):
        print("button released", b.text)
    def statetest(self,x,y):
        print("State Getting Changed")
    def build(self):
        btn1 = Button(text="Button",
```

```
            font_size=40,
            color=(1, 0, 0, 1),
            size_hint=(0, 0),
            background_color=(0, 1, 0, 1))
        btn=Button(text="Click Me",
            font_size=40,
            color=(1, 0, 0, 1),
            size_hint=(0,0),
            background_color=(0, 1, 0, 1))
        lbl = Label(text="Good Morning",
            font_size=40,
            color=(0, 1, 0, 1),
            size_hint=(0.2, 0.2))
        btn1.bind(on_press=self.printme,
            on_release=self.printed,
            state=self.statetest)
        layout=PageLayout()
        layout.add_widget(btn1)
        layout.add_widget(lbl)
        layout.add_widget(btn)
        return layout
if __name__=='__main__':
    MainApp().run()
```

Figure 8.14: Page layout.

Summary

- Layout is one UI component that instructs the child components to render in the available space. Each device is unique that makes application UI development very difficult.
- Layouts help us to control the rendering of child components. According to the way of rendering Layouts can be classified in various types. For example, BoxLayout arranges child components adjacent to the available screen space according to size_hint, pos_hint. FloatLayout arranges according to the arbitrary location of components.
- Float layout uses size_hint, pos_hint as None.
- Anchor layout arranges child components by position related to the border.
- Grid layout locates components by rectangular or grid, it is a combination of rows and columns.
- The relative layout is alike float layout, but it arranges relatively to parent layout instead of available screen space.
- Stack layout places widgets over each other in predefined size in parent widget. Page layout stacks screen on each other. These screens are scrollable from left to right or right to left. The properties like size_hint and pos_hint are not valid in page layout.

Key terms

Each device is unique that makes application UI development very difficult. According to the way of rendering, Layouts can be classified in various types. BoxLayout arranges child components adjacent to the available screen space according to size_hint, pos_hint. Float layout uses size_hint, pos_hint as None. Grid layout locates components by rectangular or grid, it's a combination of rows and columns. The relative layout is alike float layout, but it arranges relative to parent layout instead of available screen space. Stack layout places widgets over and over in predefined size in parent widget. Page layout screens are scrollable from left to right or right to left.

Review questions

1) What is layout?
2) What is the difference between float layout and relative layout?
3) What is the functionality of size_hint and pos_hint?
4) What is page layout? Explain.
5) Difference between stack and page layout.

Exercise

Tick the correct option

Q.1. Which of the essential device characteristics that we should consider when we design and develop our application _____?
a) Screen size and density
b) Devices features
c) Platform version
d) All the above

Q.2. Which of the screen size is not available _____?
a) Small
b) Normal
c) Large
d) All the available

Q.3. Layouts in Kivy are _____
a) BoxLayout
b) GridLayout
c) PageLayout
d) All of the above

Q.4. Which layout can be swapped with each other from left to right or right to left _____?
a) FloatLayout
b) PageLayout
c) BoxLayout
d) GridLayout

Q.5. Stack layout is used _____
a) to display child in a predefined size
b) to swap screen
c) to stack over and over
d) all the above

Q.6. Anchor layout position is _____
a) relative to the available screen space
b) relative to a border(left, right, top, bottom, center) of the layout
c) manually arranging widgets
d) relative to a border left
e) relative to a border(top, bottom, center) of the layout

Q.7.   In which layout child components have stacked on each other _____?
a) StackLayout
b) BoxLayout
c) GridLayout
d) PageLayout

Q.8.   Which of the following layout divides the available space between its children equally _____?
a) StackLayout
b) StackLayout
c) BoxLayout
d) GridLayout

Q.9.   Which layout allows placing their children in an arbitrary location _____?
a) BoxLayout
b) GridLayout
c) StackLayout
d) FloatLayout

Q.10. Which layout allows putting the children at a position relative to a border of the layout _____?
a) AnchorLayout
b) BoxLayout
c) StackLayout
d) GridLayout

Answers

Q.1.   d) All of the above
Q.2.   b) Normal
Q.3.   d) All of the above
Q.4.   b) PageLayout
Q.5.   a) to display child in a predefined size
Q.6.   b) relative to a border(left, right, top, bottom, center) of the layout
Q.7.   d) PageLayout
Q.8.   c) BoxLayout
Q.9.   d) FloatLayout
Q.10. a) AnchorLayout

Fill in the blanks

1. Any layout widget does not render but just acts as a _____that arranges its children in a specific way.
2. Box layout arranges widgets in an _____way to fill all the available space.
3. Page payout does not support _____.
4. _____arranges widgets to one another but with a set size in one of the dimensions, without trying to make them fit within the entire space.
5. _____allows arranging the child at a position relative to border of the layout.

Answers

1. trigger
2. adjacent
3. size_hint, size_hint_min, size_hint_max, pos_hint
4. StackLayout
5. RelativeLayout

9 Designing user interfaces

If you're talking about Java in particular, Python is about the best fit you can get among all the other languages. Yet the funny thing is, from a language point of view, JavaScript has lot in common with Python, but it is a sort of restricted subset. ~Guido Van Rossum

9.1 What is UI (user interface) components

The user interface (UI) components are different types of views that can be used to design the layout screen for the app. Basically, views are of three types:

Basic views – such as Labels, input text, buttons, image button, checkbox, toggle button, radio button, and radio group.

Picker view – that enables one to select from the list such as TimePicker and DatePicker.

List view – displays a long list of items and to select one of them, such as list view and recycler view.

9.2 Buttons

Buttons are used to perform some task by clicking on them and this action triggers while the button is pressed/released/clicked/touched. Every button has some specific properties such as border, font size, font face, font style, and padding:

```python
from kivy.app import App
from kivy.uix.button import Button
from kivy.uix.widget import Widget

class MainApp(App):
    def printme(self,a):
        print("Yes Button has been Clicked")

    def printed(self,b):
        print("button released")

    def build(self):
        btn=Button(text="Click Me",
            font_size=30,
            color=(1, 0, 0, 1),
            size_hint=(0.2,0.2),
            background_color=(0, 1, 0, 1))
```

https://doi.org/10.1515/9783110689488-009

```
        btn.bind(on_press=self.printme, on_release=self.printed)
        return btn
if __name__=='__main__':
    MainApp().run()
```

Figure 9.1: Button demo.

And the same program can be written as follows using Kv language:

```
#filename main.kv
BoxLayout:
    #size: (100, 100)
    #size_hint:(None, None)
    canvas:
      Color:
          rgb: 1,0,0
      Rectangle:
          size: (5,5)
          pos: (0,0)
cols: 2
padding : 40,40
spacing: 40, 40
row_default_height: '50dp'
```

```
background_color: 0, .5, .5, 1
orientation:"vertical"
Button:
    id:test_btn
    text_size: self.size
    font_size: 25
    text_color:1,0,0,1
    on_press: print('Button Clicked')
    #on_press: root.display_data(uname.text)
    text:"OK"
    valign: "middle"
    halign: "center"
    padding_x: 5
    background_color: 0, 1, 1, 1
    size_hint_y: None
    size_hint_x: None
    #height: self.parent.height * 0.120
    #width:self.width/2
Button:
    id:'cancelbtn'
    color:1,0,1,1
    canvas:
        Color:
            rgb: 1,1,0,1
Rectangle:
        size: (5,5)
        pos: self.pos
    text:"Close"
    size_hint_y: None
    size_hint_x: None
    on_press: app.stop()
-------------------------------MainApp-----------------------------
from kivy.app import App
from kivy.uix.button import Button
from kivy.uix.widget import Widget

class MainApp(App):
    pass

if __name__=='__main__':
    MainApp().run()
```

Figure 9.2: Button example.

9.3 Labels

The label is used to display a piece of information to the user. It acts as the label to the app, let us have a look on the given code:

```python
import kivy
from kivy.app import App
from kivy.uix.button import Button
from kivy.uix.label import Label
kivy.require("1.0.6")

class LabelTestingApp(App):
    def build(self):
        Window.clearcolor = (0, 1, 0, 1)
        return Label(text='I am Label', pos=(0,0), size_hint=(.5,.5), color=
        (0,0,1,1),font_size=40)

if __name__ == '__main__':
    LabelTestingApp().run();
```

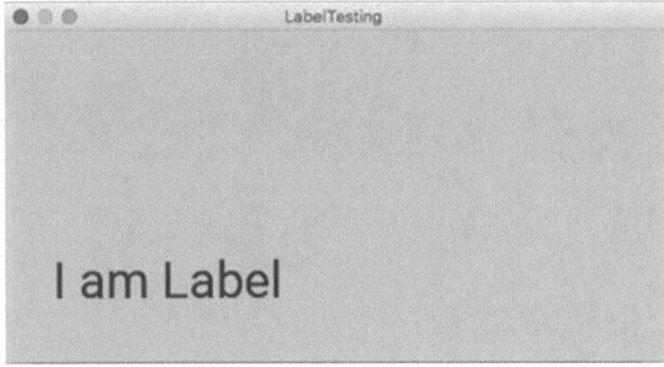

Figure 9.3: Label example.

And the same program can be written as follows using Kv language:

```
#filename main.kv
BoxLayout:
    #size: (100, 100)
    #size_hint:(None, None)
    canvas:
        Color:
            rgb: 1,0,0
        Rectangle:
            size: (5,5)
            pos: (0,0)
    cols: 2
    padding : 40,40
    spacing: 40, 40
    row_default_height: '50dp'
    background_color: 0, .5, .5, 1
    orientation:"vertical"
    Label:
        id:test_btn
        text_size: self.size
        font_size: 25
        text_color:1,0,0,1
        text:"Label 1"
        valign: "middle"
        halign: "center"
        padding_x: 5
        background_color: 0, 1, 1, 1
```

```
            size_hint_y: None
            size_hint_x: None
            #height: self.parent.height * 0.120
            #width:self.width/2
    Label:
            id:'cancelbtn'
            color:1,0,1,1
            canvas:
            Color:
                rgb: 1,1,0,1
    Rectangle:
                size: (5,5)
                pos: self.pos
    text:"Label 2"
    size_hint_y: None
  size_hint_x: None
------------------------------MainApp--------------------------------
from kivy.app import App
from kivy.uix.button import Button
from kivy.uix.widget import Widget

class MainApp(App):
```

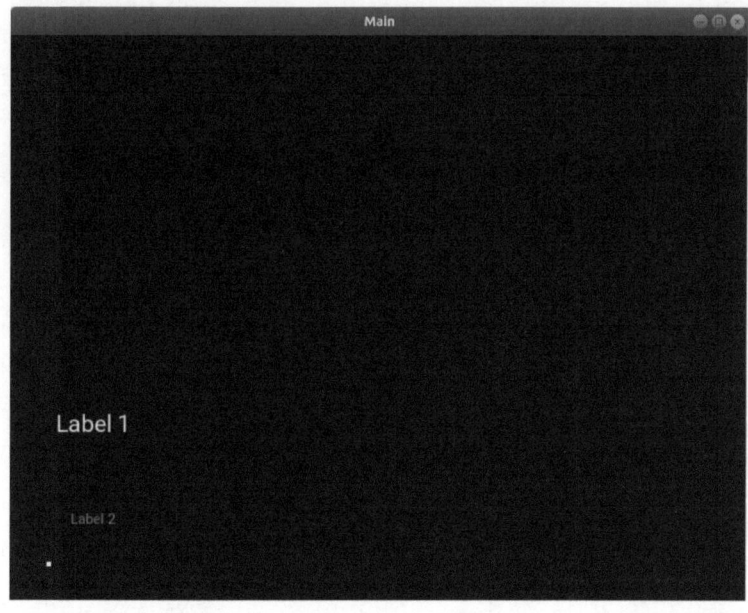

Figure 9.4: Label with KV.

```
    pass
if __name__=='__main__':
    MainApp().run()
```

9.4 ListView

ListView is a very important part in GUI programming:

```
from kivy.app import App
from kivy.lang import Builder
from kivy.uix.button import Button
items = [
{"color":(1, 1, 1, 1), "font_size": "20sp", "text": "white", "input_data":
["some","random","data"]},
{"color":(.5,1, 1, 1), "font_size": "30sp", "text": "lightblue", "input_-
data": [1,6,3]},
{"color":(.5,.5,1, 1), "font_size": "40sp", "text": "blue", "input_data":
[64,16,9]},
{"color":(.5,.5,.5,1), "font_size": "70sp", "text": "gray", "input_data":
[8766,13,6]},
{"color":(1,.5,.5, 1), "font_size": "60sp", "text": "orange", "input_data":
[9,4,6]},
{"color":(1, 1,.5, 1), "font_size": "50sp", "text": "yellow", "input_data":
[852,958,123]}
]
class MyButton(Button):
    def print_data(self,data):
        print(data)
        KV = '''
<MyButton>:
    on_release:
        root.print_data(self.input_data)
RecycleView:
    data: []
    viewclass: 'MyButton'
    RecycleBoxLayout:
        default_size_hint: 1, None
        orientation: 'vertical'
'''
class Test(App):
```

```
    def build(self):
        root = Builder.load_string(KV)
        root.data = [item for item in items]
        return root
if __name__=="__main__":
    Test().run()
```

Figure 9.5: Button demo.

9.4.1 RecyclerView

When we have to display a lot of information on the screen then we face performance issues with our device, this thing usually happens when we use ListView. To resolve this type of problem we have RecyclerView. It displays only a limited amount of records in recycling mode, meaning it will not create many objects into memory. It is memory efficient as it loads the contents into the memory according to the size of the screen while ListView loads all the items into memory that causes performance issues. We can apply animations on recycle view to create an interactive list. It has two major components **View Holder** and **Adapter**. View Holder is responsible to hold the view and recycle it while the adapter is responsible to hold data into memory to load on the recycle view holder. It provides a flexible model for viewing selected data. Its design is based on model view control. Let us see the given code:

```
from random import sample
from string import ascii_lowercase
```

```python
from kivy.app import App
from kivy.lang import Builder
from kivy.uix.boxlayout import BoxLayout

kv = """
<Row@BoxLayout>:
    canvas.before:
        Color:
            rgba: 0.5, 0.5, 0.5, 1
        Rectangle:
            size: self.size
            pos: self.pos
    value: ''
    Label:
        text: root.value
<Test>:
    canvas:
        Color:
            rgba: 0.3, 0.3, 0.3, 1
        Rectangle:
            size: self.size
            pos: self.pos
    rv: rv
    orientation: 'vertical'
    GridLayout:
        cols: 3
        rows: 2
        size_hint_y: None
        height: dp(108)
        padding: dp(8)
        spacing: dp(16)
        Button:
            color:0,1,0,1
            font_size:35
            background_color:1,0,0,1
            text: 'Populate list'
            on_press: root.populate()
        Button:
            color:0,0,1,1
            font_size:35
            background_color:1,0,0,1
            text: 'Sort list'
```

```
                on_press: root.sort()
            Button:
                color:0,0,0,1
                font_size:35
                background_color:1,1,.2,1
                text: 'Clear list'
                on_press: root.clear()
            BoxLayout:
                spacing: dp(8)
                Button:
                    color:0,1,0,1
                    font_size:15
                    background_color:1,0,0,1
                    text: 'Insert new item'
                    on_press: root.insert(new_item_input.text)
                TextInput:
                    id: new_item_input
                    size_hint_x: 0.6
                    hint_text: 'value'
                    padding: dp(10), dp(10), 0, 0
            BoxLayout:
                spacing: dp(8)
                Button:
                    text: 'Update first item'
                    on_press: root.update(update_item_input.text)
                TextInput:
                    id: update_item_input
                    size_hint_x: 0.6
                    hint_text: 'new value'
                    padding: dp(10), dp(10), 0, 0
            Button:
                text: 'Remove first item'
                color:1,0,0,1
                font_size:15
                background_color:1,0,0,1
                on_press: root.remove()
    RecycleView:
        color:1,.5,.5,1
        id: rv
        scroll_type: ['bars', 'content']
        scroll_wheel_distance: dp(114)
        bar_width: dp(10)
```

```
            viewclass: 'Row'
            RecycleBoxLayout:
                default_size: None, dp(56)
                default_size_hint: 1, None
                size_hint_y: None
                size_hint_x: 1
                height: self.minimum_height
                #width:self.minimum_width
                orientation: 'vertical'
                spacing: dp(6)
"""
Builder.load_string(kv)

class Test(BoxLayout):
    def populate(self):
        self.rv.data = [{'value': ''.join(sample(ascii_lowercase, 20))} for
        x in range(100)]

    def sort(self):
        self.rv.data = sorted(self.rv.data, key=lambda x: x['value'])

    def clear(self):
        self.rv.data = []

    def insert(self, value):
        self.rv.data.insert(0, {'value': value or 'No Data'})

    def update(self, value):
        if self.rv.data:
            self.rv.data[0]['value'] = value or 'No New Data'
            self.rv.refresh_from_data()

    def remove(self):
        if self.rv.data:
            self.rv.data.pop(0)

class TestApp(App):
    def build(self):
        return Test()

if __name__ == '__main__':
    TestApp().run()
```

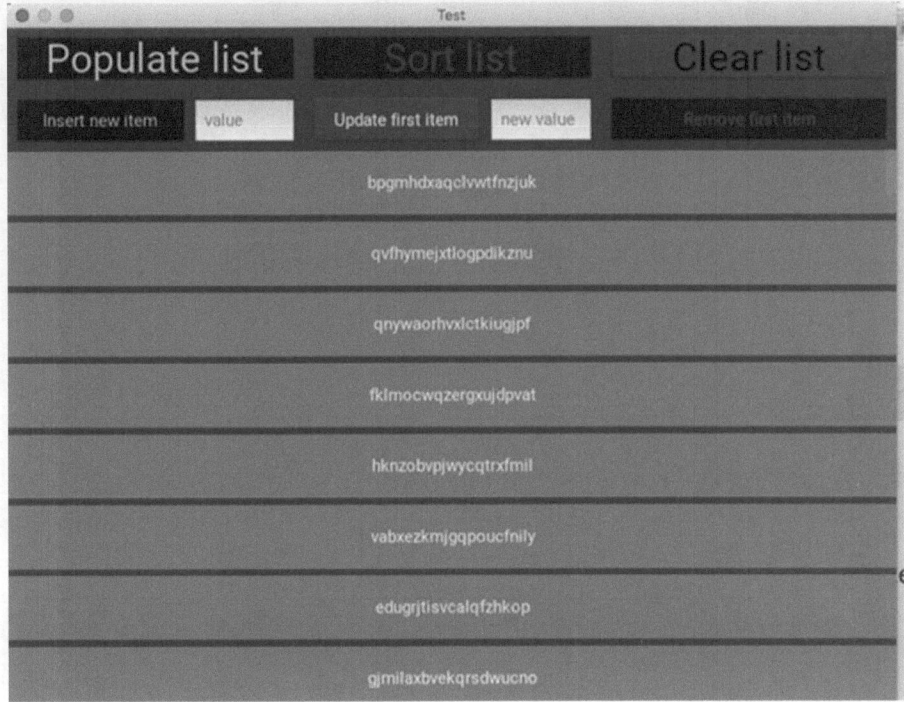

Figure 9.6: Recycle view.

9.5 Image

In the earlier topics we have learned about various views that are used to display information in various ways, and some views are used to collect information from the user. If we want to show an image on the screen, Image view is used:

```python
from kivy.app import App
from kivy.uix.image import Image
from kivy.loader import Loader

class TestApp(App):
    def _image_loaded(self, proxyImage):
        if proxyImage.image.texture:
            self.image.texture = proxyImage.image.texture

    def build(self):
        proxyImage = Loader.image("9.png")
        proxyImage.bind(on_load=self._image_loaded)
        self.image = Image()
```

```
        return self.image

TestApp().run()
```

Figure 9.7: Image display.

9.6 Events and properties

Properties make it easy to pass updates from Python objects to GUI objects while binding passes the changes from GUI to Python objects. Let us see in the given example:

```
--------------------------------main.kv--------------------------------
#filename main.kv
BoxLayout:
    size: (500, 500)
    size_hint:(None,None)
    canvas:
        Color:
            rgb: 1,0,0
        Rectangle:
            size: (5,5)
            pos: (0,0)
```

```
    Widget:
        id:wig
        pos: (250,250)
        canvas:
            Color:
                rgb: 1,1,1
            Rectangle:
                size: (5,5)
                pos: self.pos
    Label:
        id: boo
        text:"boo"
        color: 0,0,1,1
        size_hint:(1,1)
        pos_hint:{"center_x":1,"center_y":1}
    Label:
        id: foo
        text: "foo"
        color: 0,1,0,1
        size_hint: (.6,.6)
        pos_hint:{"x":1,"y":1}
    Label:
        id: bar
        text:"bar"
        color: 1,0,0,1
        size:(500,500)
        size_hint:(None,None)
        pos_hint:{"right":1,"top":1}
        pos:100, 10
    Button:
        text:"goo"
        size_hint:0.1,0.1
        pos:(1,1)
        #some debug info, i know the code is ugly
on_press: print (self.parent.size,'\n', self.parent.right, self.parent.top,
self.parent.x, self.parent.y, self.parent.center_x, self.parent.center_y,
"\n","bar_right_top:", bar.pos,"foo_x_y:", foo.pos,"boo_center:", boo.pos,
"\nwhite square:", wig.pos, "\n", bar.size, foo.size, boo.size)
```

```python
from kivy.app import App
from kivy.uix.widget import Widget
```

```
class MainApp(App):
    pass
if __name__=="__main__":
    MainApp().run()
```

Figure 9.8: Events.

9.7 Fonts and their properties

By default, Kivy supports a number of fonts, namely, Roboto-Bold, Roboto-BoldItalic, Roboto-Itelic, Roboto-Regular, RobotoMono-Regular, and DejaVuSans. The list can be found in the URL: https://github.com/kivy/kivy/tree/master/kivy/data/fonts. Font is a namespace where multiple fonts are loaded. If a font is missing a glyph needed to render text, it can fall back to a different font in the same context. The font context manager can be used to query and manipulate the state of font context while using the **Pango** text provider. Font context can be created automatically by widget.

9.8 Popup menu

Popup menu displays the menu below the anchor text if space is available, otherwise it appears above the anchor text. It disappears if we click outside the popup menu. Let us see the given code:

```python
from kivy.app import App
from kivy.uix.popup import Popup
from kivy.lang import Builder
from kivy.uix.button import Button

Builder.load_string('''
<SimpleButton>:
    on_press: self.fire_popup()
<SimplePopup>:
    id:pop
    size_hint: .4, .4
    auto_dismiss: False
    title: 'Hello world!!'
    Button:
        text: 'Click here to dismiss'
        on_press: pop.dismiss()
''')

class SimplePopup(Popup):
    pass

class SimpleButton(Button):
    text = "Fire Popup !"
    def fire_popup(self):
        pops=SimplePopup()
        pops.open()

class SampleApp(App):
    def build(self):
        return SimpleButton()

SampleApp().run()
```

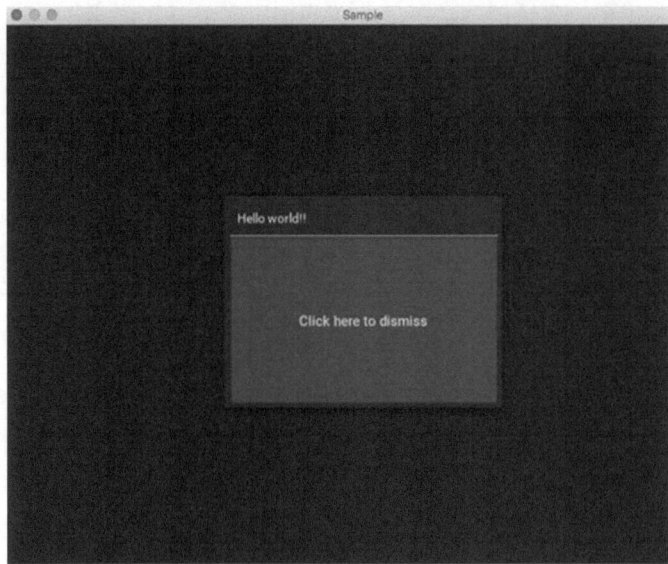

Figure 9.9: Popup menu.

9.9 TextInput

TextInput is used to get input in the Python program. Let us see the code:

```python
from kivy.app import App
from kivy.uix.button import Button
from kivy.uix.textinput import TextInput
from kivy.uix.boxlayout import BoxLayout

class ClearApp(App):
    def build(self):
        self.box = BoxLayout(orientation='vertical', spacing=20)
        self.txt = TextInput(hint_text='Your text goes here', size_hint=(1,
1))
        self.btn = Button(text='Clear All', on_press=self.clearText, size_-
hint=(1, 1))
        self.box.add_widget(self.txt)
        self.box.add_widget(self.btn)
        return self.box

def clearText(self, instance):
        self.txt.text = ''
ClearApp().run()
```

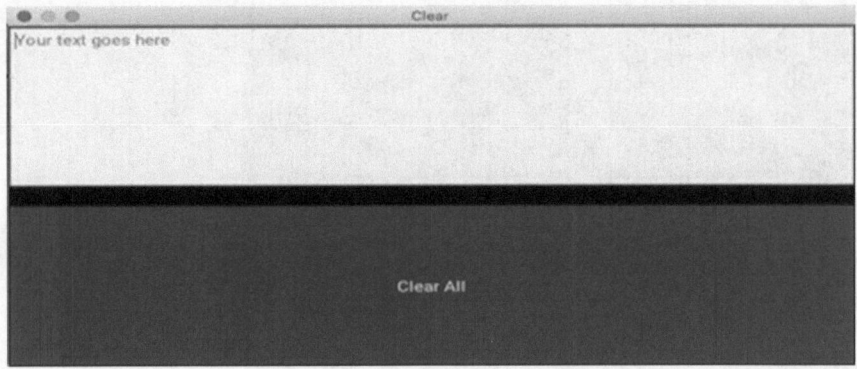

Figure 9.10: Text input.

9.10 ScrollView

The ScrollView provides a scrollable view. It can be scrolled on both x axis and y axis on the screen. Function **runTouchApp()** is responsible to scrollview touch. Let us see in the code.

```
from kivy.base import runTouchApp
from kivy.lang import Builder
root = Builder.load_string(r'''
ScrollView:
    Label:
        text: 'Mastering Mobile Application Development using Python with
        Kivy' * 100
        font_size: 30
        size_hint_x: 1.0
        size_hint_y: None
        text_size: self.width, None
        height: self.texture_size[1]
''')

runTouchApp(root)
```

Figure 9.11: Scroll view.

9.11 Kivy clock

The Kivy clock object is used to schedule a function call after a specific period or interval. Let us see the given code:

```python
from kivy.app import App
from kivy.uix.button import Button
from kivy.clock import Clock

class ClockExample(App):
    i = 0
    def build(self):
        self.mybtn = Button(text='Number of Calls')
        Clock.schedule_interval(self.clock_callback, 2)
        return self.mybtn
    def clock_callback(self, dt):
        self.i += 1
        self.mybtn.text = "Call = %d" % self.i
ClockExample().run()
```

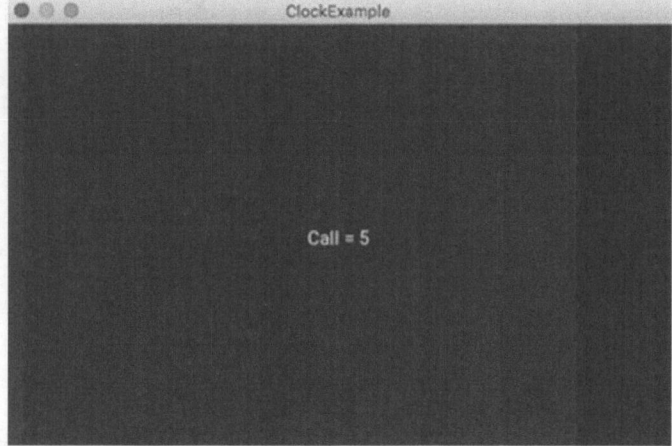

Figure 9.12: Kivy clock.

Summary

- UI components are views that are used to design responsive and attractive app. These are of three types: basic view, picker view, and list view.
- Basic views are labels, radio buttons, radio group, input text, buttons, and so on.
- Picker views are date picker and time picker.
- List view is used to display long list, such as ListView and RecyclerView.
- Button is used to trigger some events when it is pressed, released, clicked, or touched. A button has properties like padding, font face, font style, font size, and border.
- Label is the way to display information on the app.
- ListView and Recycler View are used to display the long list of data. Recycler View is the best choice when we have a long list to display.
- Image is used to display any image as asynchronously or synchronously. Every component has some events and properties to render on the screen.
- Event binding is very important to communicate between Python objects to UI objects.
- Kivy supports many fonts. A font is a namespace where multiple fonts are loaded. If a font is missing a glyph needed to render text, it automatically falls back to the different font in the same context.
- Popup menu is used to anchor text, and it is very eager to display text even when screen space not available.
- User can provide input text from the input text.
- Scrollview can scroll on both x and y axes.
- Kivy clock is used to schedule a function call after certain period or interval.

Key terms

Kivy UI components are basic, picker, and list views. Any component of UI fires events when it is being clicked, touched, pressed, and so on, and these events are needed to bind with the components. Binding is the way to communicate between Python objects to GUI components. Kivy clock is responsible to perform some tasks at a certain time and period. UI scrollable property comes from the scroll view. This scrolling is available on both axes x and y. The popup menu is eager UI that can be rendered on the priority over the screen. Kivy supports various fonts to display text from its loaded context.

Review questions

1. What types of UI components are available in Kivy?
2. How to load context works to render text in a different font?
3. What is the difference between ListView and RecyclerView?
4. How Image and AsyncImage are different?
5. What is event binding in Kivy, and how is it useful in Kivy?
6. What is the difference between basic and picker views?

Exercise

Tick the correct option

Q.1. Which is not a type of Kivy component _____?
   a) Button
   b) List
   c) Picker
   d) Basic view

Q.2. Which of the following is a list view _____?
   a) Button view
   b) Time picker view
   c) Recycler view
   d) CheckBox view

Q.3. Which of these is a type of label _____?
   a) List View
   b) Basic view
   c) Picker view
   d) None of the above

Q.4. Which view is loaded in image view_____?
a) Asynchronously
b) Synchronously
c) Do not know
d) All of the above

Q.5. Which is not a button property_____?
a) Font style
b) Padding
c) Border
d) None of the above

Q.6. Fonts are loaded from Kivy namespace_____
a) single font
b) multiple font
c) only selected font
d) None of the above

Q.7. How event binding is important to Kivy_____?
a) To make Kivy program easy
b) To improve application performance
c) To communicate between Python object and GUI components
d) To protect from virus

Q.8. How is Kivy clock useful_____?
a) To call set date and time to Kivy application
b) To call a function after a certain period or interval
c) To connect the database on time
d) To connect server

Q.9. Can the scroll view be scrolled on_____?
a) x axis
b) y axis
c) z axis
d) Both a and b

Q.10. Which of the UI component can help to get input text from
keyboard_____?
a) Button
b) Checkbox
c) Input text
d) Radio button

Answers

Q.1. a) Button
Q.2. c) RecyclerView
Q.3. b) Basic view
Q.4. b) synchronously
Q.5. d) None of the above
Q.6. b) multiple
Q.7. c) to communicate between Python object and GUI components
Q.8. b) to call a function after a certain period or interval
Q.9. d) Both a and b
Q.10. c) Input text

Fill in the blanks

1. UI components are views that are used to design_____ app.
1. The main stakeholders of the cloud ecosystem are the _____.
2. There are two types of list views, namely_____ and _____.
3. A button has properties like _____.
4. Image view loads image _____ while asyncImage loads_____.
5. Popup menu is used to_____ even when space not available.

Answers

1. responsive and attractive
2. ListView, RecyclerView
3. padding and font face border
4. synchronously, asynchronously
5. anchor text

10 UX widgets

My Favorite language for maintainability is Python. It has simple clean syntax, object encapsulation, good library support, and optional named parameters.   ~Bram Cohen

10.1 ActionBar and slider

ActionBar was introduced for the first time in the 1.8.0 version of Kivy. It is like
Android's ActionBar, where items are stacked horizontally or vertically. It contains
one ActionView and many ContextualActionView. An ActionView will contain an
ActionPrevious that has title, app_icon, previous_icon properties. An ActionView contains ActionItems, some of them are ActionButton, ActionToggleButton, ActionCheck,
ActionSeparator, and Action Group. When an area becomes too small, widgets are
moved into the ActionOverFlow area.

ActionGroup is used to show ActionItems in the group. ActionView shows an
ActionGroup after other ActionItem. ActionView contains ActionOverflow.
ContextualAction is a subclass of ActionView:

```python
from kivy.base import runTouchApp
from kivy.factory import Factory as F

actionbar = F.ActionBar(pos_hint={'top': 1})

av = F.ActionView(background_color= (1, .5, 0, 1))
av.add_widget(F.ActionPrevious(title='My Action Bar', with_previous=False))
av.add_widget(F.ActionOverflow())
av.add_widget(F.ActionButton(text='Btn0',
        icon='atlas://data/images/defaulttheme/audio-volume-low'))

for i in range(1, 10):
  av.add_widget(F.ActionButton(text='Btn{}'.format(i)))

ag = F.ActionGroup(text='Group1')
for i in range(11, 15):
  ag.add_widget(F.ActionButton(text='Btn{}'.format(i)))

av.add_widget(ag)

actionbar.add_widget(av)

# can't be set in F.ActionView() -- seems like a bug
av.use_separator = True

runTouchApp(actionbar)
```

https://doi.org/10.1515/9783110689488-010

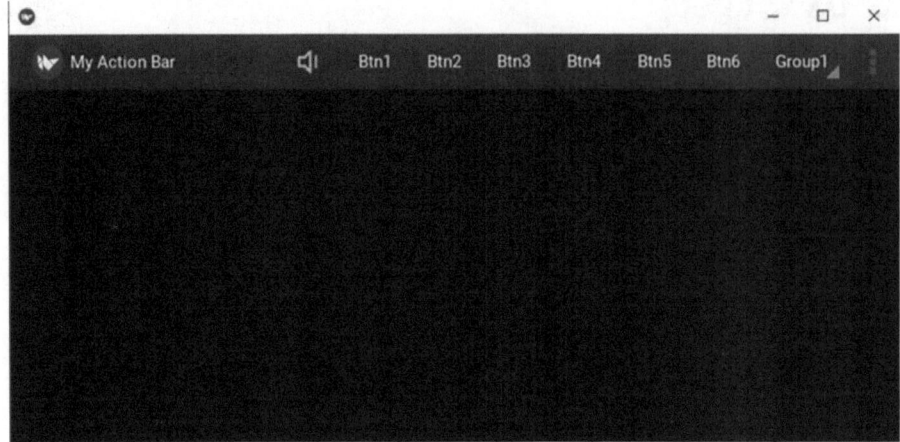

Figure 10.1: ActionBar demo.

Slider is a scroll bar. It supports horizontal and vertical orientations, min/max values, and a default value. Kivy supports several slider options for customizing between the minimum value and maximum value. Kivy also supports several slider options for customizing the cursor, cursor image, border, and background to be used in different orientations and regions between the minimum value and the maximum value:

```python
from kivy.app import App
from kivy.core.window import Window
from kivy.uix.boxlayout import BoxLayout
from kivy.uix.button import Button
from kivy.uix.slider import Slider
from kivy.uix.widget import Widget

class SliderExample(App):
    def build(self):
        root=BoxLayout()
        Window.clearcolor = (.5, .5, 1, 1)
        #root = Widget()
        root.add_widget(Button(text="Click Me"))
        slider = Slider()
        root.add_widget(slider)
        #root.clear_widgets()
        #remove_widget(slider)
        return root
if __name__=='__main__':
    SliderExample().run()
```

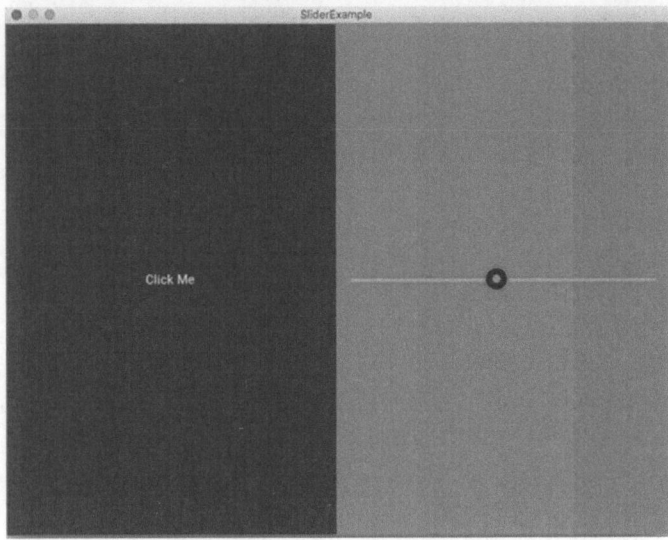

Figure 10.2: Button and slider.

10.2 Checkbox and text on window

Checkbox is a unique two-state button that can be either checked or unchecked. It provides single and multiple check options. If used in the group, it becomes radio button that ensures single selection within available options (see Figure 10.3):

Figure 10.3: Checkbox window.

The checkbox is the part of kivi.uix package, so it can be used in code like the given code:

```
from kivy.uix.label import Label
from kivy.uix.checkbox import CheckBox
from kivy.uix.gridlayout import GridLayout
from kivy.graphics import InstructionGroup
from kivy.graphics import Color
from kivy.graphics import Rectangle
from kivy.app import App
from kivy.core.window import Window
class UserInterface(GridLayout):
```

```python
    def __init__(self, **kwargs):
        super(UserInterface, self).__init__(**kwargs)
        # Add labels and checkbox
        self.cols = 2
        self.add_widget(Label(text='Accept License'))
        self.cb_roateMode = CheckBox(active=True)
        self.add_widget(self.cb_roateMode)
        self.lbl_roateMode = Label(text='License Accepted')
        self.add_widget(self.lbl_roateMode)
        # Change the color of the background to grey
        backgroundColor = InstructionGroup()
        backgroundColor.add(Color(1, .5, 1, .3))
        backgroundColor.add(Rectangle(pos=self.pos,     size=(Window.width,
        Window.height)))
        self.canvas.add(backgroundColor)
        # Attach a callback
        self.cb_roateMode.bind(active=self.on_roateMode_Active)
        # Callback for the checkbox
def on_roateMode_Active(self, checkboxInstance, isActive):
    if isActive:
        self.lbl_roateMode.text = "License Accepted"
    else:
        self.lbl_roateMode.text = "License not Accepted"
# App derived from App class
class CheckBoxApp(App):
    def build(self):
        ux = UserInterface()
        return ux
# Run the app
if __name__ == '__main__':
    CheckBoxApp().run()
```

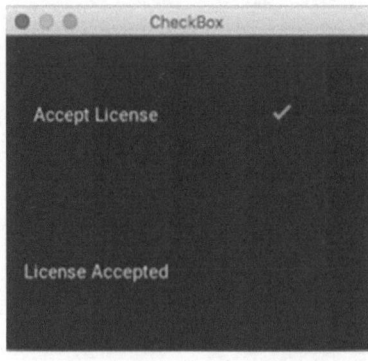

Figure 10.4: Checkbox demo.

Text box is used to get input from user; let us see the code:

```
#:kivy 1.10.0
<CustomLabel@Label>:
    text_size: self.size
    valign: "middle"
    padding_x: 5
<SingleLineTextInput@TextInput>:
    multiline: False
<GreenButton@Button>:
    background_color: 1, 1, 1, 1
    size_hint_y: None
    height: self.parent.height * 0.120
UserGroup
    GridLayout:
        cols: 2
        padding : 30,30
        spacing: 20, 20
        row_default_height: '30dp'
        CustomLabel:
            text: 'Name'
            text_size: self.size
            valign: 'middle'
        SingleLineTextInput:
id: uname
        CustomLabel:
            text: 'Address'
            text_size: self.size
            valign: 'middle'
        SingleLineTextInput:
            id: uaddress
        CustomLabel:
            text: 'Phone'
            text_size: self.size
            valign: 'middle'
        SingleLineTextInput:
            id: uphone
        GreenButton:
            text: 'Ok'
            on_press:  root.display_data(uname.text,uaddress.text,uphone.
text)
```

```
        GreenButton:
            text: 'Cancel'
            on_press: app.stop()
        Label:
            text: ''
            text_size: self.size
        valign: 'middle'
```

```
-------------------------------MainApp-------------------------------
from kivy.uix.label import Label
from kivy.uix.popup import Popup
from kivy.uix.screenmanager import Screen
from kivy.app import App
from kivy.lang import Builder
from kivy.core.window import Window
from kivy.properties import StringProperty

Window.size = (400, 300)
class UserGroup(Screen):
    gender = StringProperty("")
    def display_data(self, uname,uaddress,uphone):
        popup = Popup(title="User Information",
                content=Label(text=uname+"\n"+uaddress+"\n"+uphone),
                auto_dismiss=False)
        popup.open()
class FactUserGroup(App):
    def build(self):
        self.root = Builder.load_file('test.kv')
        return self.root

if __name__ == '__main__':
    FactUserGroup().run()
```

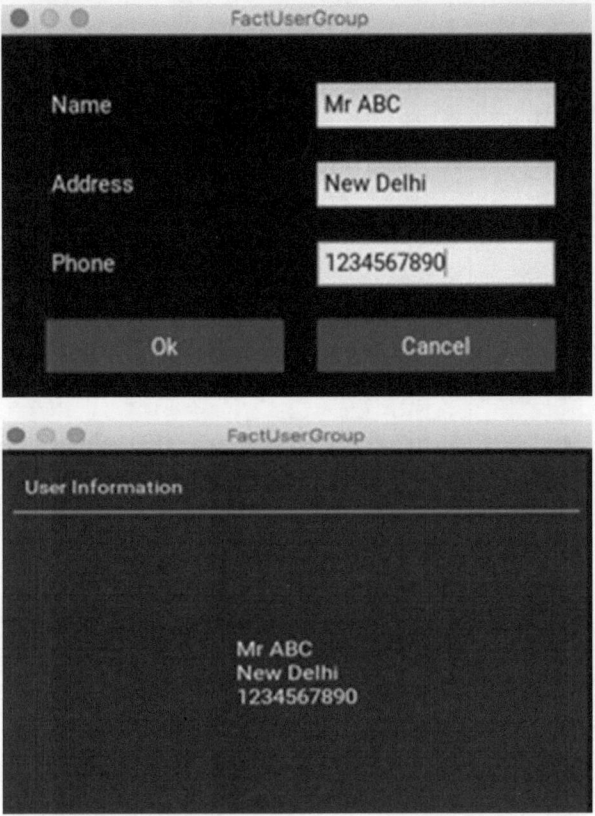

Figure 10.5: Text input example.

10.2.1 Toggle button

Toggle button behaves like a checkbox, the only difference is when we click or touch, it changes its state normal to down. Toggle buttons can also be grouped to make radio buttons – only one button in a group can be in a downstate. A group name can be a string or hash-able Python object. Only one choice can be selected at a time.

Figure 10.6: Toggle button.

Let us see the following code:

```
#from ipywidgets import ToggleButton
from kivy.app import App
from kivy.uix.anchorlayout import AnchorLayout
```

```python
from kivy.uix.button import Button
from kivy.uix.floatlayout import FloatLayout
#import prompt_toolkit.formatted_text
from kivy.uix.gridlayout import GridLayout
from kivy.uix.label import Label
from kivy.uix.textinput import TextInput
from kivy.uix.togglebutton import ToggleButton

class MyApp(App):
    def build(self):
        layout = GridLayout()
        layout.padding=1
        layout.cols=2
        btn = Button(text='Submit')
        layout.add_widget(btn)
        canclebtn = Button(text='Cancle')
        Testbtn = Button(text='Test Button')
        layout.add_widget(canclebtn)
        layout.add_widget(Testbtn)
        parentlayout=GridLayout()
        parentlayout.rows=4
        upperlayout=GridLayout()
        upperlayout.rows=7
        name=TextInput()
        name.hint_text="Student Name only"
        label=Label()
        label.text="Student Name"
        address = TextInput()
        address.hint_text = "Student Address only"
        addresslabel = Label()
        addresslabel.text = "Student Address"
        phone = TextInput()
        phone.hint_text = "Student Phone only"
        phonelabel = Label()
        phonelabel.text = "Student Phone"
        btn1 = ToggleButton(text='Male', group='sex',)
        btn2 = ToggleButton(text='Female', group='sex', state='down')
        btn3 = ToggleButton(text='Third Gender', group='sex')
        lowerLayout=GridLayout()
        lowerLayout.cols=4
        gender = Label()
        gender.text = "Gender"
```

```
        lowerLayout.add_widget(gender)
        lowerLayout.add_widget(btn1)
        lowerLayout.add_widget(btn2)
        lowerLayout.add_widget(btn3)
        upperlayout.add_widget(label)
        upperlayout.add_widget(name)
        upperlayout.add_widget(addresslabel)
        upperlayout.add_widget(address)
        upperlayout.add_widget(phonelabel)
        upperlayout.add_widget(phone)
        upperlayout.add_widget(lowerLayout)
        #upperlayout.add_widget(layout)
        parentlayout.add_widget(upperlayout)
        #parentlayout.add_widget(lowerLayout)
        parentlayout.add_widget(layout)
        return parentlayout
if __name__=='__main__':
    MyApp().run()
```

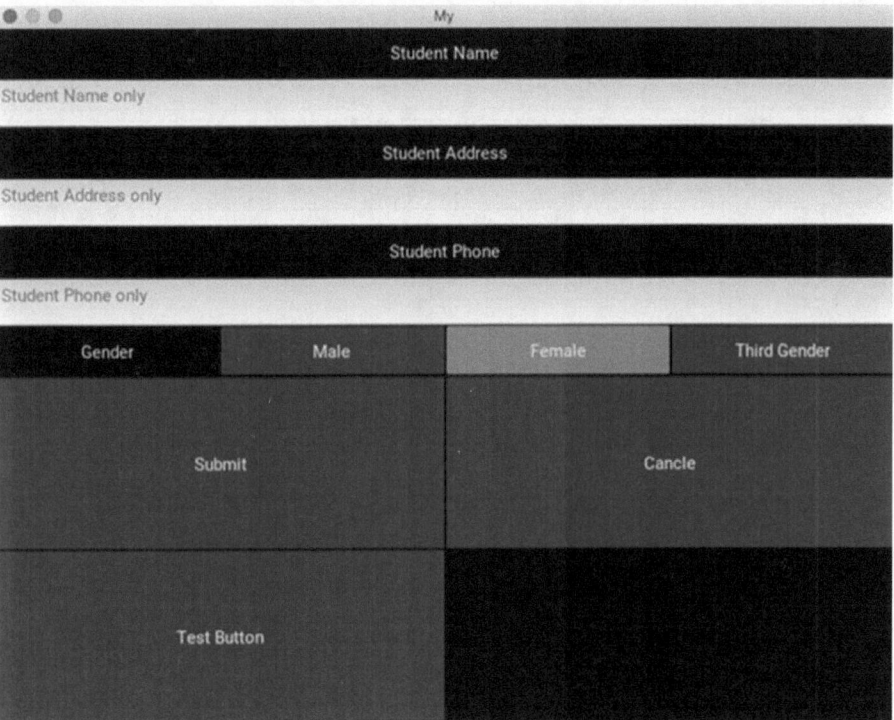

Figure 10.7: Toggle button example.

10.2.2 TreeView

TreeView is a widget that is used to represent a tree structure. The TreeView always creates a default root node, based on TreeViewLabel. A TreeView is populated with TreeViewNode instances, but we cannot use a TreeViewNode directly. We have to combine it with widgets such as Label and Button. It is currently very basic, supporting a minimal feature set:

```python
from kivy.app import App
from kivy.uix.button import Button
from kivy.uix.gridlayout import GridLayout
from kivy.uix.scrollview import ScrollView
from kivy.uix.treeview import TreeView, TreeViewLabel

class MyApp(App):
    def build(self):
        tv = TreeView()
        tv.add_node(TreeViewLabel(text='My first item'))
        tv = TreeView()
        oneYear = tv.add_node(TreeViewLabel(text='One Year Program'))
        twoYear = tv.add_node(TreeViewLabel(text='Two Year Program'))
        sixMonth = tv.add_node(TreeViewLabel(text='Six Month Program'))
        threeMonth = tv.add_node(TreeViewLabel(text='Three Month Program'))
        tv.add_node(TreeViewLabel(text='Data Science'), oneYear)
        tv.add_node(TreeViewLabel(text='Data Anylytics'), oneYear)
        tv.add_node(TreeViewLabel(text='CCNA'), oneYear)
        tv.add_node(TreeViewLabel(text='Python Programming'), threeMonth)
        tv.add_node(TreeViewLabel(text='JavaScript           Programming'),
        threeMonth)
        tv.add_node(TreeViewLabel(text='Mean Stack'), threeMonth)
        tv.add_node(TreeViewLabel(text='Web Desine'), sixMonth)
        tv.add_node(TreeViewLabel(text='Web Development'), sixMonth)
        tv.add_node(TreeViewLabel(text='Graphics Design'), sixMonth)
        networking=tv.add_node(TreeViewLabel(text='Networking'),twoYear)
        tv.add_node(TreeViewLabel(text='Semester-1'), networking)
        tv.add_node(TreeViewLabel(text='Semester-2'), networking)
        tv.add_node(TreeViewLabel(text='Semester-3'), networking)
        tv.add_node(TreeViewLabel(text='Semester-4'), networking)
        tv.add_node(TreeViewLabel(text='Semester-5'), networking)
```

```
software = tv.add_node(TreeViewLabel(text='Software'),twoYear)
swsemeste1=tv.add_node(TreeViewLabel(text='Semester-1'), software)
swsemeste2=tv.add_node(TreeViewLabel(text='Semester-2'), software)
swsemeste3=tv.add_node(TreeViewLabel(text='Semester-3'), software)
swsemeste4=tv.add_node(TreeViewLabel(text='Semester-4'), software)
swsemeste5=tv.add_node(TreeViewLabel(text='Semester-5'), software)
tv.add_node(TreeViewLabel(text='Working Smart with Office and
Internet'), swsemeste1)
tv.add_node(TreeViewLabel(text='Logic Building'), swsemeste1)
tv.add_node(TreeViewLabel(text='SQL Server 2012'), swsemeste1)
tv.add_node(TreeViewLabel(text='Advanced Excel'), swsemeste1)
tv.add_node(TreeViewLabel(text='Programming in Java'), swsemeste2)
tv.add_node(TreeViewLabel(text='HTML5 CSS and JavaScript '),
swsemeste2)
tv.add_node(TreeViewLabel(text='JSP and Servlet'), swsemeste2)
tv.add_node(TreeViewLabel(text='Android'), swsemeste3)
tv.add_node(TreeViewLabel(text='Hibernate Spring and JSF'),
swsemeste3)
tv.add_node(TreeViewLabel(text='JUnit'), swsemeste3)
tv.add_node(TreeViewLabel(text='C# Programming'), swsemeste4)
tv.add_node(TreeViewLabel(text='ASP.Net programming'), swsemeste4)
tv.add_node(TreeViewLabel(text='Cross Plateform Applocation By
HTML-5 CSS and JavaScript'), swsemeste4)
tv.add_node(TreeViewLabel(text='Distribute               Application
Development'), swsemeste5)
tv.add_node(TreeViewLabel(text='ORM'), swsemeste5)
tv.add_node(TreeViewLabel(text='Windows App Development using HTML5
CSS and JavaScript'), swsemeste5)
return tv

if __name__=='__main__':
    MyApp().run()
```

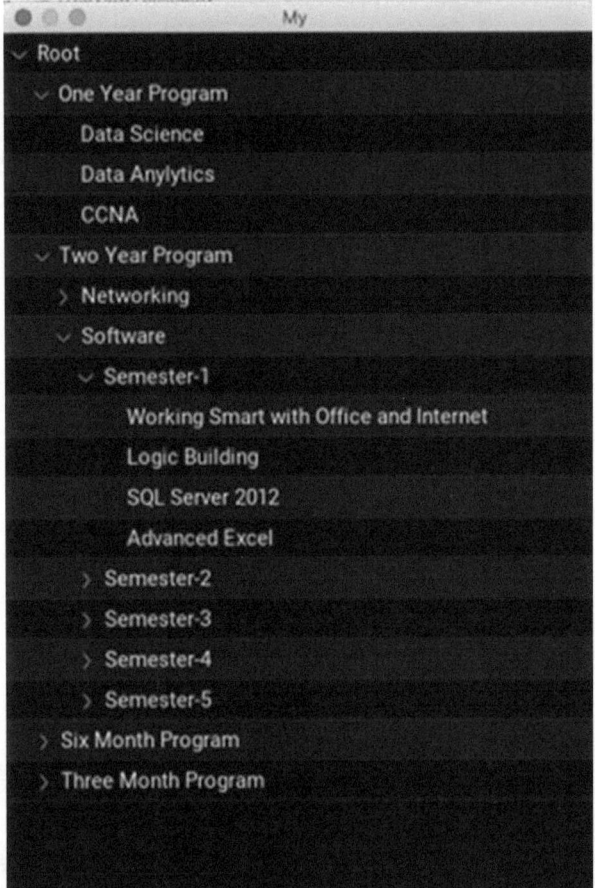

Figure 10.8: TreeView demo.

10.3 Radio/checkbox button and label management

A checkbox is a unique two-state button that can be either checked or unchecked. It provides single and multiple check options, if used in the group it becomes a radio button that ensures single selection within available options, as can be seen in the snapshot:

```
---------------------------------test.kv---------------------------------
#:kivy 1.10.0 filename test.kv
<CustomLabel@Label>:
    text_size: self.size
    valign: "middle"
    padding_x: 5
```

```
<SingleLineTextInput@TextInput>:
    multiline: False
<GreenButton@Button>:
    background_color: 1, 1, 1, 1
    size_hint_y: None
    height: self.parent.height * 0.120
UserGroup
    GridLayout:
        cols: 2
        padding : 30,30
        spacing: 20, 20
        row_default_height: '30dp'
        CustomLabel:
            text: 'Name'
            text_size: self.size
            valign: 'middle'
        SingleLineTextInput:
            id: uname
        Label:
            text: 'Male'
            text_size: self.size
            valign: 'middle'
        CheckBox:
            group: 'check'
            id : chk
            text: "Male"
            on_active:
                root.gender = self.text
        Label:
            text: 'Female'
            text_size: self.size
            valign: 'middle'
        CheckBox:
            group: 'check'
            text: "Female"
            on_active:
                root.gender = self.text
        GreenButton:
            text: 'Ok'
            on_press: root.display_data(uname.text)
        GreenButton:
            text: 'Cancel'
```

```
                on_press: app.stop()
        Label:
                text: ''
                text_size: self.size
valign: 'middle'
```

```
-----------------------------------mainapp--------------------------
from kivy.uix.label import Label
from kivy.uix.popup import Popup
from kivy.uix.screenmanager import Screen
from kivy.app import App
from kivy.lang import Builder
from kivy.core.window import Window
from kivy.properties import StringProperty

Window.size = (400, 300)
class UserGroup(Screen):
    gender = StringProperty("")
    def display_data(self, uname):
        popup   =   Popup(title=uname,   content=Label(text=self.gender),
        auto_dismiss=False)
        popup.open()

class FactUserGroup(App):
    def build(self):
        self.root = Builder.load_file('test.kv')
        return self.root

if __name__ == '__main__':
    FactUserGroup().run()
```

Labels are used to display information over the screen. It supports ASCII and Unicode strings too. The size of Label is not affected by the text content and the text is not affected by the size. In order to control the size, we must specify text_size to constrain the text. Let us see the following code:

```
from kivy.app import App
from kivy.compat import unichr
from kivy.uix.boxlayout import BoxLayout
from kivy.uix.button import Button
from kivy.uix.label import Label

class MainMenu(BoxLayout):
    def __init__(self, **kwargs):
```

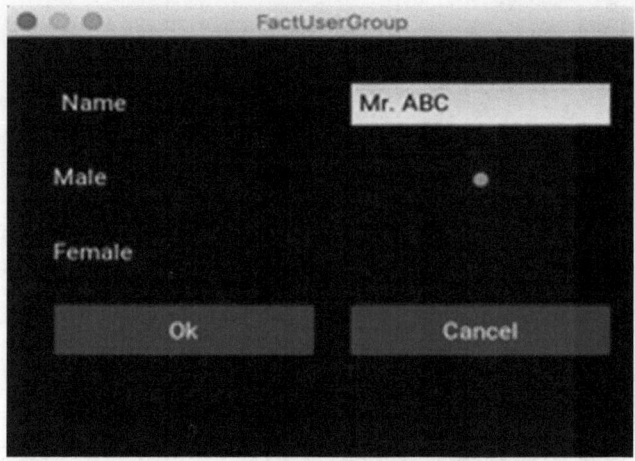

Figure 10.9: Radio button and text box.

```
        super(MainMenu, self).__init__(**kwargs)
        self.orientation = 'vertical'
        self.background_color=(0,1,0,1)
        lb2 = Label(text=u"This is [font=Arial]Unicode[/font] String with
        markup\ n"+unichr(178), markup=True, font_size="30sp", color=
        (0.5,0,1,1),shorten=True)
        lbl = Label(text="[color=ff3322]'This is [b]Bold Markup[/b]
        String'[/color]", markup=True,font_size="40sp")
        self.add_widget(Label(text='This is Ordinary ANSI UTF-8 String'))
        self.add_widget(lbl)
        self.add_widget(lb2)

class Test1App(App):
    def build(self):
        return MainMenu()
if __name__ == "__main__":
    Test1App().run()
```

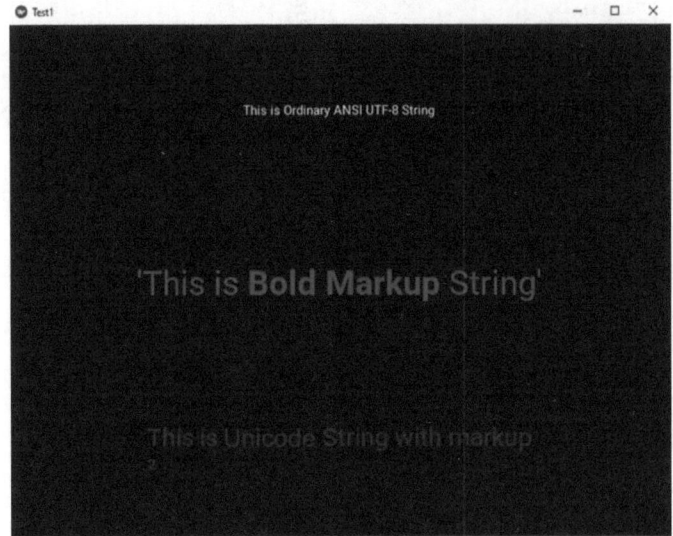

Figure 10.10: Display information using label.

10.4 ProgressBar and carousel

The progress bar is used to check the progress of completion with a certain task. The Progressbar does not have interactive elements and is a display-only widget. We can visualize the progress of some tasks. Currently, only horizontal mode is supported:

```python
from kivy.app import App
from kivy.uix.widget import Widget
from kivy.uix.button import Button
from kivy.uix.progressbar import ProgressBar
from kivy.uix.boxlayout import BoxLayout
from kivy.clock import Clock
from time import sleep

class MainMenu(BoxLayout):
    def __init__(self, **kwargs):
        super(MainMenu, self).__init__(**kwargs)
        self.orientation = 'vertical'
        btn = Button(text="Start")
        btn.bind(on_release=self.trigger)
```

```python
        self.add_widget(btn)
        self.MyList = ('this', 'is', 'a', 'test')
        self.i = 0
        self.pb = ProgressBar(max=len(self.MyList), value=0)
        self.add_widget(self.pb)
    def trigger(self, *args):
        self.i = 0
        self.pb.value = 0
        Clock.schedule_interval(self.heavyFunc, 0.1)
    def heavyFunc(self, dt):
        sleep(0.5)
        print(self.MyList[self.i])
        self.i += 1
        self.pb.value += 1
        if self.i >= len(self.MyList):
            Clock.unschedule(self.heavyFunc)
            print('unscheduled')
class Test1App(App):
    def build(self):
        return MainMenu()
if __name__ == "__main__":
    Test1App().run()
```

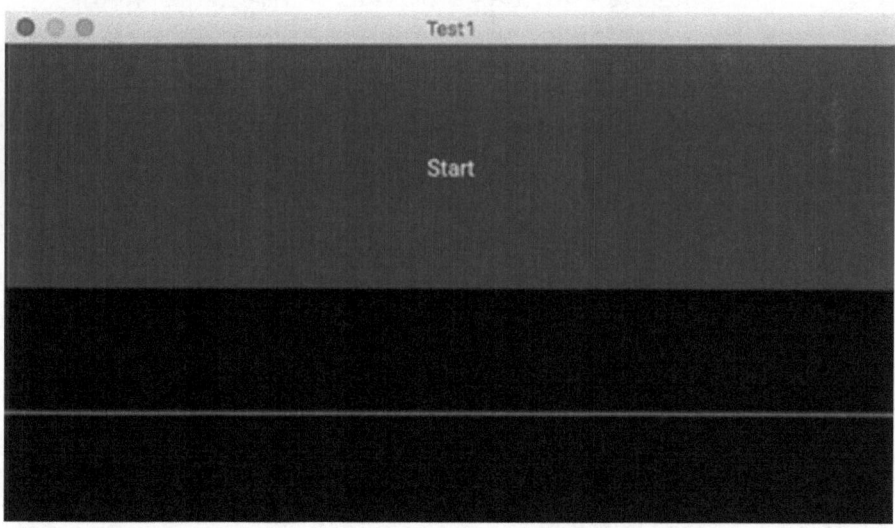

Figure 10.11: Carousel example.

Kivy clock

Kivy clock allows us to schedule a function call in the future. This function call may be scheduled once or repeatedly at specified intervals. We can get the time elapsed between the scheduling and the calling of the callback via *dt* argument. If the callback returns False, the schedule will be canceled and never repeated:

```python
import kivy
from kivy.app import App
from kivy.clock import Clock
from kivy.uix.button import Button
from kivy.uix.popup import Popup
from kivy.uix.progressbar import ProgressBar
from kivy.uix.widget import Widget
from kivy.properties import ObjectProperty
kivy.require('1.5.1')

class MyWidget(Widget):
    progress_bar = ObjectProperty()
    def __init__(self, **kwa):
        super(MyWidget, self).__init__(**kwa)
        self.progress_bar = ProgressBar()
        self.popup = Popup(title='Popup', content=self.progress_bar)
        self.popup.bind(on_open=self.puopen)
        self.add_widget(Button(text='Download', on_release=self.pop))

def pop(self, instance):
        self.progress_bar.value = 1
        self.popup.open()

def next(self, dt):
        if self.progress_bar.value>=100:
            return False
        self.progress_bar.value += 1

def puopen(self, instance):
        Clock.schedule_interval(self.next, 1/25)

class MyApp(App):
        def build(self):
            return MyWidget()

if __name__ in ("__main__"):
MyApp().run()
```

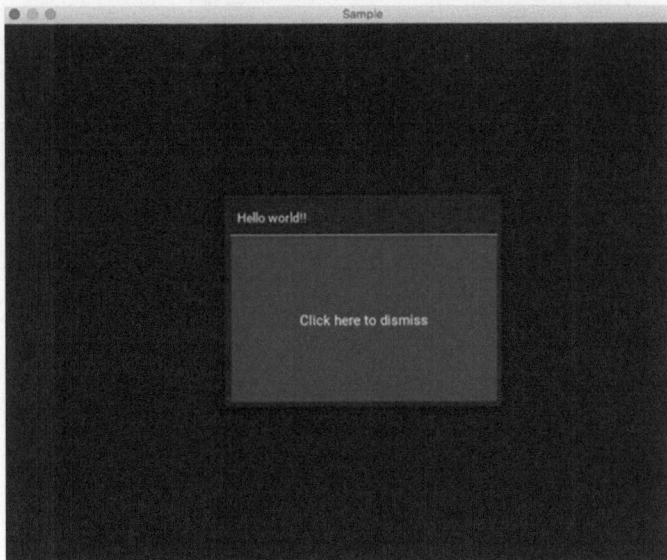

Figure 10.12: Kivy clock.

10.4.1 Custom ProgressBar

To customize a ProgressBar, it requires defining the attribute or properties for the background and progress of our progress bar. We need to extend Python class as a child class of ProgressBar class, and then child class will occupy all the properties of the parent class. In the child class, we can make the required changes that are not needed from the parent class. We can do the same thing with kv language by extending child class. There are two types of progress bars – the determinate progress bar (for a fixed duration) and the indeterminate progress bar (for unknown duration):

```python
from kivy.app import App
from kivy.uix.progressbar import ProgressBar
from kivy.core.text import Label as CoreLabel
from kivy.lang.builder import Builder
from kivy.graphics import Color, Ellipse, Rectangle
from kivy.clock import Clock
class CircularProgressBar(ProgressBar):
    def __init__(self, **kwargs):
        super(CircularProgressBar, self).__init__(**kwargs)
```

```
                # Set constant for the bar thickness
                self.thickness = 60
                # Create a direct text representation
                self.label = CoreLabel(text="0%", font_size=self.thickness)
                # Initialise the texture_size variable
                self.texture_size = None
                # Refresh the text
                self.refresh_text()
                # Redraw on innit
                self.draw()
    def draw(self):
        with self.canvas:
                # Empty canvas instructions
                self.canvas.clear()
                # Draw no-progress circle
                Color(0.5, 0.50, 0.26)
                Ellipse(pos=self.pos, size=self.size)
                # Draw progress circle, small hack if there is no progress (an-
                gle_end = 0 results in full progress)
                Color(1, 1, 0)
                Ellipse(pos=self.pos, size=self.size,
                        angle_end=(0.001 if self.value_normalized == 0 else
                        self.value_normalized*360))
                # Draw the inner circle (colour should be equal to the background)
                Color(0, 0, 0)
                Ellipse(pos=(self.pos[0] + self.thickness / 2, self.pos[1] + self.
                thickness / 2),
                        size=(self.size[0] - self.thickness, self.size[1] -
self.thickness))
                # Center and draw the progress text
                Color(1, 1, 1, 1)
                Rectangle(texture=self.label.texture, size=self.texture_size,
                        pos=(self.size[0]/2 - self.texture_size[0]/2, self.
size[1]/2 - self.texture_size[1]/2))
        def refresh_text(self):
                # Render the label
                self.label.refresh()
                # Set the texture size each refresh
                self.texture_size = list(self.label.texture.size)
```

```
    def set_value(self, value):
        # Update the progress bar value
        self.value = value
        # Update textual value and refresh the texture
        self.label.text = str(int(self.value_normalized*100)) + "%"
        self.refresh_text()
        # Draw all the elements
        self.draw()
class Main(App):
    # Simple animation to show the circular progress bar in action
    def animate(self, dt):
        if self.root.value < 80:
            self.root.set_value(self.root.value + 1)
        else:
            self.root.set_value(0)
            # Simple layout for easy example
        def build(self):
    container = Builder.load_string(
            '''CircularProgressBar:
        size_hint: (1, 1)
        height: 200
        width: 400
        max: 50''')
                # Animate the progress bar
                Clock.schedule_interval(self.animate, 0.1)
                return container
if __name__ == '__main__':
Main().run()
```

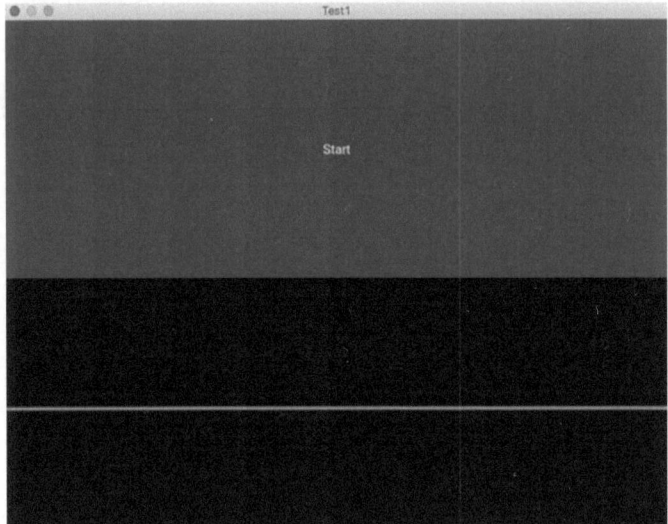

Figure 10.13: Progress bar.

10.4.2 Carousel layout

Carousel can be used to create a slide show. Here, we will have a 1-second delay between each screen. Carousel widget is the superclass of the root. We are using the clock as a timer. Root is based on the carousel. The update function runs the carousel load next function to load the next slide:

```
from kivy.uix.carousel import Carousel
from kivy.uix.gridlayout import GridLayout
from kivy.app import App
from kivy.lang import Builder
Builder.load_string('''
<Page>:
    cols: 3
    Label:
        text: str(id(root))
Button
Button
Button
Button
    text: 'load(page 3)'
    on_release:
        carousel = root.parent.parent
```

```
                carousel.load_slide(carousel.slides[2])
Button
Button
    text: 'prev'
    on_release:
        root.parent.parent.load_previous()
Button
Button
    text: 'next'
    on_release:
        root.parent.parent.load_next()
''')
class Page(GridLayout):
    pass
class TestApp(App):
    def build(self):
        root = Carousel()
        for x in range(10):
            root.add_widget(Page())
        return root
if __name__ == '__main__':
    TestApp().run()
```

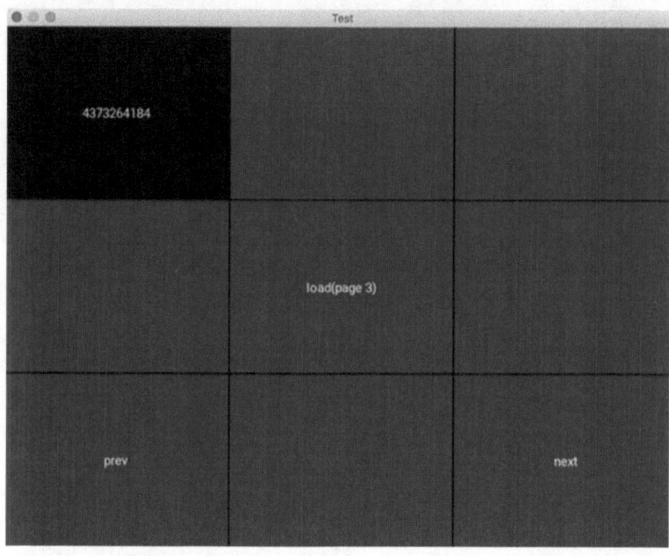

Figure 10.14: Carousel.

All the given code in this book are valid for both Python versions but, for the best results, tested. Let us see another example of Carousel view:

```python
from kivy.app import App
from kivy.uix.carousel import Carousel
from kivy.uix.image import AsyncImage

class CarouselDemo(App):
    def build(self):
        carouselObject = Carousel(direction='left')
        for i in range(10):
    imageFile = "pic.png"
    asyncImageObject = AsyncImage(source=imageFile)
    carouselObject.add_widget(asyncImageObject)
carouselObject.index = 2
return carouselObject

# Start the Carousel App
if __name__ == '__main__':
    CarouselDemo().run()
```

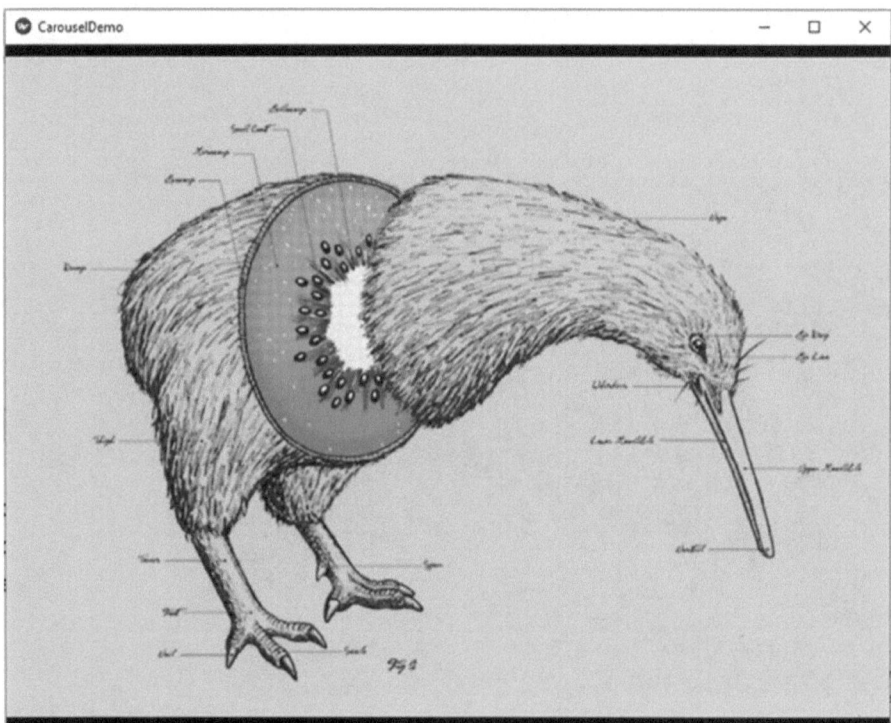

Figure 10.15: AsyncImage with Carousel.

10.4.3 Scatter

Scatter provides feature to drag and drop over the screen. Scatter has its own matrix transformation. It is used to build interactive widgets that can be translated, rotated, and scaled with two or more fingers on a multitouch system. Children's components are positioned relative to scatter similarly. Let us see how we can achieve it:

```python
from kivy.app import App
from kivy.properties import StringProperty
from kivy.lang import Builder

KV = """
#:import win kivy.core.window
<Picture@Scatter>:
    source: None
    on_size: self.center = win.Window.center
    size: image.size
    size_hint: None, None
    Image:
        id: image
        source: root.source
FloatLayout:
    Picture:
        source: "img/2.png"
    Picture:
        source: "img/1.png"
"""
class MyApp(App):
    def build(self):
        return Builder.load_string(KV)
MyApp().run()
```

Figure 10.16: Scatter view.

Let us see another example for drag and drop objects:

```python
from kivy.app import App
from kivy.uix.scatter import Scatter
from kivy.uix.label import Label
from kivy.uix.floatlayout import FloatLayout

class TutorialApp(App):
    def build(self):
        f = FloatLayout()
        s = Scatter()
        l = Label(text="Drag and Drop Me",
        font_size=150)
        f.add_widget(s)
        s.add_widget(l)
        return f
if __name__ == "__main__":
    TutorialApp().run()
```

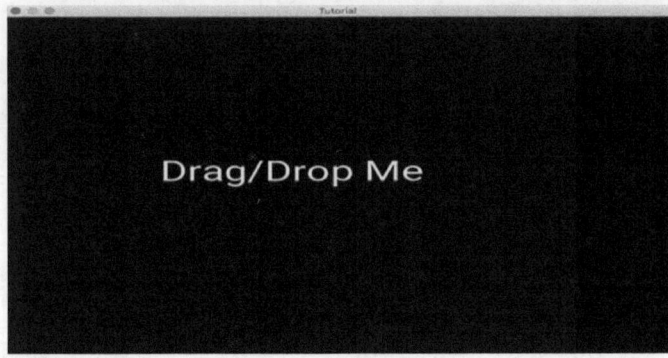

Figure 10.17: Drag and drop.

10.5 Canvas

Canvas is the root object for drawing by a widget. It uses the performance of the graphics engine by drawing a large number of small dots. Canvas is available to all widgets. When we create a widget, we can create all the instructions required for drawing. For more information, we can refer to the documentation:

```
from kivy.app import App
from kivy.lang import Builder
kv = '''
BoxLayout:
    orientation: 'vertical'
    BoxLayout:
        size_hint_y: None
        height: sp(100)
        BoxLayout:
            orientation: 'vertical'
            Slider:
                id: e1
                min: -360.
                max: 360.
            Label:
                text: 'angle_start = {}'.format(e1.value)
        BoxLayout:
            orientation: 'vertical'
            Slider:
                id: e2
                min: -360.
                max: 360.
```

```
                    value: 360
                Label:
                    text: 'angle_end = {}'.format(e2.value)
        BoxLayout:
            size_hint_y: None
            height: sp(100)
            BoxLayout:
                orientation: 'vertical'
                Slider:
                    id: wm
                    min: 0
                    max: 2
                    value: 1
                Label:
                    text: 'Width mult. = {}'.format(wm.value)
            BoxLayout:
                orientation: 'vertical'
                Slider:
                    id: hm
                    min: 0
                    max: 2
                    value: 1
                Label:
                    text: 'Height mult. = {}'.format(hm.value)
            Button:
                text: 'Reset ratios'
                on_press: wm.value = 1; hm.value = 1
        FloatLayout:
            canvas:
                Color:
                    rgb: 1, 1, 1
                Ellipse:
                    pos: 100, 100
                    size: 200 * wm.value, 201 * hm.value
                    source: '9.png'
                    angle_start: e1.value
                    angle_end: e2.value
'''
class CircleApp(App):
def build(self):
return Builder.load_string(kv)
CircleApp().run()
```

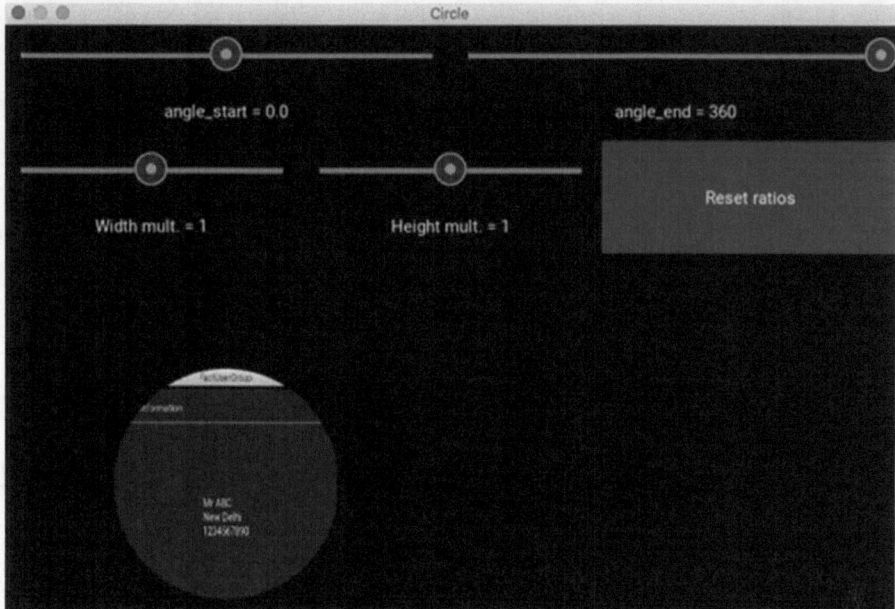

Figure 10.18: Canvas.

We must keep in mind a few things before using canvas:
1) Canvas is not the space in which we paint like HTML5 canvas.
2) It is a set of instructions not like bitmap image(.bmp) or vector images (GIF, PNG images).
3) Widgets share the same coordinate space, not the same canvas. No special space is being provided to the canvas.
4) Color is an exception because of PushMatrix and PopMatrix do not have an effect over the Color.
5) The instruction execution order follows a traversal order.
6) Drawing before the children with canvas.before.
7) Drawing after the children with canvas.after.

10.6 AsyncImage

Image and AsyncImage both are used to display an image over the screen. AsyncImage is used to load image asynchronously, and it is useful as it prevents our application from waiting until the image is loaded. If we want to display a large image or retrieve them from URLs, using AsyncImage will allow these resources to be retrieved on a background thread without blocking our application, and image by default is centered and fits inside the widget binding box. If we do not want that we can set allow_stretch to True and keep_ratio to False, we can inherit from Image and create our own style, as shown in the following code:

```python
from kivy.app import App
from kivy.uix.image import AsyncImage
from kivy.lang import Builder

Builder.load_string('''
<CenteredAsyncImage>:
    size_hint: 0.8, 0.8
    pos_hint: {'center_x': 0.5, 'center_y': 0.5}
    mipmap: True
''')
class CenteredAsyncImage(AsyncImage):
    pass
class TestAsyncApp(App):
    def build(self):
        url = ('9.png')
        return CenteredAsyncImage(source=url)

if __name__ == '__main__':
    TestAsyncApp().run()
```

Figure 10.19: AsyncImage view.

10.7 Spinner

Spinner is alternate to the radio button. It provides a quick way to select one value from a given set. It displays only a selected value at a time. A spinner can have multiple values to be selected. Spinner updates its text properties every time attr:values are changed:

```python
from kivy.app import App
from kivy.core.window import Window
from kivy.uix.boxlayout import BoxLayout
from kivy.uix.button import Button
from kivy.uix.slider import Slider
from kivy.uix.widget import Widget

class SliderExample(App):
    def build(self):
        root=BoxLayout()
        Window.clearcolor = (.5, .5, 1, 1)
        #root = Widget()
        root.add_widget(Button(text="Click Me"))
        slider = Slider()
        root.add_widget(slider)
        #root.clear_widgets()
        #remove_widget(slider)
        return root

if __name__=='__main__':
    SliderExample().run()
```

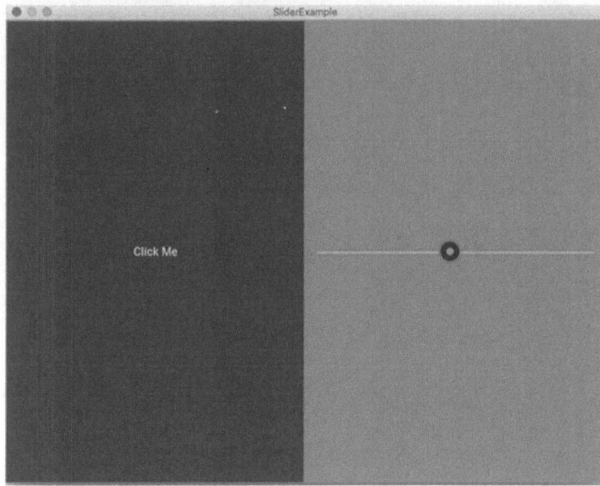

Figure 10.20: Slider demo.

10.8 Accordion

An accordion is a form of the menu where the options are stacked either vertically or horizontally, and the item click opens to display its content. It may contain plenty of items with individual root content widgets. It has two types – first one container for the title bar and second another container for the content. Let us have a look at the example:

```python
from kivy.app import App
from kivy.uix.accordion import Accordion

class Accor(Accordion):
    pass

class AccorApp(App):
    def build(self):
        return Accor()

if __name__ == '__main__':
    AccorApp().run()
```

Figure 10.21: Accordion.

```
# accor.kv file of the Accordion App file

<MyImage@Image>:
    keep_ratio: False
    allow_stretch: True

<Accor>:
    orientation: 'horizontal'
    AccordionItem:
        title: 'Chestnut'
        MyImage:
            source: 'pic/chestnut.webp'
    AccordionItem:
        title: 'Chilly/Pepe'
        MyImage:
            source: 'pic/chilly.webp'
    AccordionItem:
        title: 'Coffee Beans'
        MyImage:
            source: 'pic/coffee.webp'
    AccordionItem:
        title: 'Fruit'
        MyImage:
            source: 'pic/fruit.webp'
    AccordionItem:
        title: 'Grains'
        MyImage:
            source: 'pic/grains.webp'
```

10.9 Switching between two screens

Now, we know that we have multiple activities in our app. But these activities have no use if they are not related or linked to each other. In this section, we will see how these activities can be linked to each other. Linking two activities means that we want to pass information from one activity to another activity by click or simple pass true or false value. In our app, we have two screens and we are switching between them, as shown in the code:

```
import kivy
kivy.require('1.9.0')
from kivy.app import App
from kivy.lang import Builder
```

```
from kivy.uix.screenmanager import ScreenManager, Screen

# You can create your kv code in the Python file
Builder.load_string("""
<FirstScreen>:
    BoxLayout:
        Button:
            text: "Click or Swap for Second Screen"
            on_press:
                # You can define the duration of the change and the direction
                of the slide
                root.manager.transition.direction = 'left'
                root.manager.transition.duration = 1
                root.manager.current = 'second_screen'
<SecondScreen>:
    BoxLayout:
        Button:
            text: "Click or Swap for First Screen"
            on_press:
                root.manager.transition.direction = 'right'
                root.manager.current = 'first_screen'
""")
# Create a class for all screens in which you can include
# helpful methods specific to that screen
class FirstScreen(Screen):
    pass

class SecondScreen(Screen):
    pass

# The ScreenManager controls moving between screens
screen_manager = ScreenManager()
# Add the screens to the manager and then supply a name
# that is used to switch screens
screen_manager.add_widget(FirstScreen(name="first_screen"))
screen_manager.add_widget(SecondScreen(name="second_screen"))
class SwitchingScreenApp(App):
    def build(self):
        return screen_manager

sample_app = SwitchingScreenApp()
sample_app.run()
```

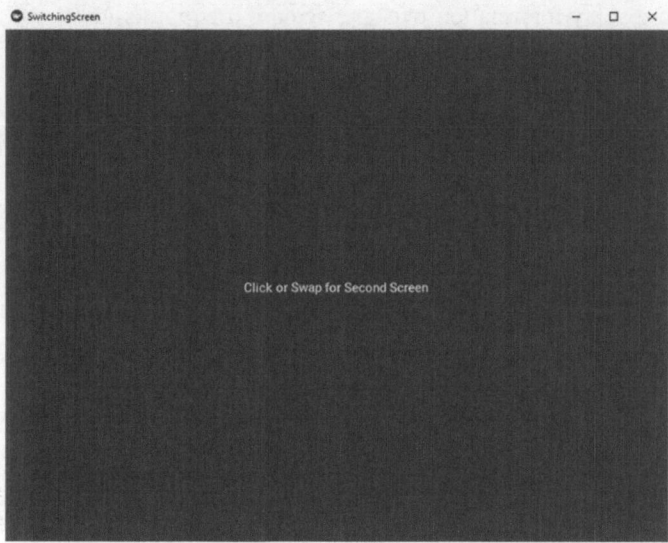

Figure 10.22: Swapping screens.

Summary

- ActionBar is like ActionBar in Android where items are stacked horizontally. ActionView contains ActonItems, ActionButtons, ActionToggle Buttons, ActionCheck, ActionSeperator and ActionGroup. It was introduced in Kivy 1.8.0 for the first time.
- Slider is a horizontal or vertical scrollbar. It supports minimum/maximum value and default value.
- Checkbox is a special two-state button that can be either checked or unchecked. It provides single (if we are using in the group) and multiple check options (without any group). Group name can be a string or a hash-able Python object.
- Toggle Button is like checkbox the only difference is when we click or touch, it changes its state normal to down. It can be used in a group to convert into radio buttons where it ensures only one button can be down at a time.
- The TreeView is used to represent a tree structure. It always creates a default root node, which is based on TreeViewLabel. A TreeView is populated with TreeViewNode instances, but we cannot use a TreeViewNode directly. We need to combine it with widgets like Label and Button.
- Checkbox is a special two-state button that can be either checked or unchecked. It provides single and multiple check options. If we use it in a group, it becomes a radio button that ensures single selection within available options.

- Labels are used to display information over the screen, it supports ASCII and Unicode strings too. The size of the label is not affected by text content and the text is not affected by the size.
- The progress bar is used to check the progress of completion with a certain task. The progress bar does not have interactive elements and is a display-only widgets. We can visualize the progress of some tasks. Currently, it supports only horizontal orientation.
- The Kivy clock allows us to schedule a function call in the future. This function call can be scheduled maybe for once or repeatedly at specified intervals. We can get the time elapsed between the scheduling and the calling of the callback via *dt* argument. If a callback function returns False, it means schedule has been canceled and will never be repeated.
- The progress bar requires defining the attribute and properties for the background and progress of our progress bar. We need to extend Python class as a child class of progress bar class then child class will occupy all the properties from the parent class. In the child class, we can make the required changes that are not needed from the parent class. The progress bars are of two types – determinate progress bar and indeterminate progress bar.
- The Carousel is used to create a slide show. Carousel widget is a superclass of the root that is based on the carousel. Update function runs the carousel load next function to load the next slide.
- Scatter is used to drag and drop features over screen. Scatter has its own matrix transformation. It is used to build interactive widgets that can be translated, rotated, and scaled with two or more fingers on a multitouch system.
- The canvas is the root object for drawing by a widget. It is not the space in which we can paint like HTML5 or other canvas. It uses the performance of the graphics engine by drawing a large number of small dots. Canvas is available to all widgets. Canvas is a set of instructions in Kivy. All widgets share the same coordinate space, not the same canvas, means no special space will be allocated to canvas.
- The Image and asyncImage are used to display image over screen. AysncImage is used to load Image asynchronously. It is useful as it prevents our application from waiting till the image is loaded. If we want to display a large image or retrieve from URLs, using AsyncImage will allow these resources to be retrieved on background thread without blocking our application. By default, image is centered and fits inside the widget boundary box. We can avoid this default behavior by setting *allow_stretch=True*
- The spinner is an alternate of the radio button. It provides a quick way to select one value from a given set. It displays only a selected value at a time. A spinner can have multiple values to be selected. Spinner updates its text properties every time attr:value is changed.

- The accordion is a form of menu where the options are stacked either vertically or horizontally. The item click opens to display its content. It may contain plenty of items with individual root content widget. It has two types – first one container for the title bar and second another container for the content.
- More than one activity can be linked to each other. It requires to pass information from one activity to another activity by click or touch. This information passes through the activities as True or False.

Key terms

The action bar was introduced in Kivy 1.8.0. Action bar contains ActionItems, ActionButton, ActionToggleButtons, ActionCheck, ActionSeperator, and ActionGroup. Slider supports orientation horizontal and vertical. A checkbox is a special state button that can be either checked or unchecked. If the checkbox is being used along with group, then it turns into a radio button. The checkbox groupname can be a string or hash-able Python object. Toggle Button is one of them with some different feature that supports only on/off state. While used in the group, it allows only one selection at a time. TreeView always creates a default root node, which is based on TreeViewLabel. A TreeView is populated with TreeViewNode instances but we cannot use a TreeViewNode directly. We need to combine it with widgets such as labels and buttons. Labels are used to display ASCCI and Unicode base string over the screen. The progress bar is used to check the progress of completion of the task. The progress bar has no interactive elements. Currently, the progress bar supports only horizontal orientation. Progress bar requires defining the attribute and properties for the background and progress of our progress bar. Kivy clock allows us to schedule a function call in the future. This function may be called once or repeatedly at a certain period. The carousel is used to create a slide show. The scatter is used to drag and drop features over screen. The scatter has its own matrix transformation, which is used to build interactive widgets that can be translated, rotated, and scaled with two or more fingers on a multitouch screen. The canvas is the root object for drawing by a widget. It is not the space in which we paint like HTML5. The Image and AsyncImage are used to display images over the screen. AsyncImage is more efficient than Image because it waits for image loading from URL. Loading image from URL is a very time-consuming task and still, there is no guarantee that the image loads successfully. This process is introduced by another new child thread, it avoids pausing user thread. Spinner is an alternate to the radio button. It provides a quick way to select one value from a given set. Spinner can allow multiple values to be selected. An accordion is a form of the menu where the options are stacked either vertically or horizontally.

Review questions

1. What is ActionBar and what are its key components?
2. How is slider different from a spinner?
3. What is the major difference between the Checkbox and radio button?
4. What is the difference between toggle button and radio button?
5. How do we create a TextInput code?
6. What is TreeView and how is it useful in designing an app?
7. What is label in Kivy and how do we use ASCII and Unicode in it?
8. How does carousel work? Explain.
9. What is the progress bar and how is it different from slider?
10. What is the use of *dt* argument in Kivy clock?
11. How do we create a custom progress bar?
12. What is the scatter view?
13. How is canvas different from HTML5 canvas?
14. What is the difference between Image and AsyncImage?
15. How is Spinner better than radio button and toggle button?

Exercise

Tick the correct option

Q.1. What is the file extension created for Kivy programming _____?
    a) ky file
    b) kb file
    c) ki file
    d) kv file

Q.2. What is alternate of radio button and toggle button _____?
    a) Spinner
    b) Checkbox
    c) Radio button
    d) Toggle button

Q.3. How many value selections can be made in toggle button _____?
    a) One
    b) Many
    c) Depends upon the usage implementation in code
    d) Do not know

Q.4. Accordion view supports which orientation _____?
    a) Horizontal
    b) Vertical
    c) Horizontal and vertical
    d) Neither horizontal nor vertical

Q.5. Which version of kivy had introduced Action bar _____?
    a) 1.7.0
    b) 1.8.0
    c) 1.8.5
    d) 1.8.4

Q.6. A group name can be _____
    a) any string
    b) any hash-able object
    c) both
    d) none of the above

Q.7. TreeView always creates a default _____
    a) TreeViewNode
    b) TreeViewLabel
    c) root node
    d) None of the above

Q.8. If we are using checkbox in a group, then it behaves like _____
    a) toggle button
    b) TextInput box
    c) Checkbox
    d) radio button

Q.9. text_size attribute is used to _____.
    a) support ASCII code
    b) control size of the text
    c) Size of the label does not get affected by text content
    d) support Unicode

Q.10. The update function runs the carousel loads next function to
    load_____.
    a) first slide
    b) last slide
    c) previous slide
    d) next slide

Answers

Q.1. d) kv file
Q.2. a) Spinner
Q.3. c) Depends upon the usage implementation in code
Q.4. c) Horizontal and vertical
Q.5. b) 1.8.0
Q.6. c) both
Q.7. c) default root node
Q.8. d) radio button
Q.9. b) control size of the text
Q.10. d) next slide

Fill in the blanks

1. Action bar is like_____ action bar.
2. The Kivy supports several slider options for customizing between the _____.
3. _____is a special two-state button that can be either checked or unchecked.
4. _____changes its state normal to down.
5. TreeView is a widget that is used to represent _____.

Answers

1. Android's
2. minimum value and maximum
3. Checkbox
4. Toggle button
5. Cloud resellers

11 Graphics handling

> In many ways, it's a dull language, borrowing solid old concepts from many other languages and styles: boring syntax, unsurprising semantics, few automatic corrections etc. But that's one of the things I like about Python.
>
> ~Tim Peters

11.1 Graphics

Using graphics programming we can make our app interactive and attractive. It requires lots of programming efforts. But when we use Python it becomes very easy as Python provides *graphics* module. In this example, we are going to draw the Sierpinski triangle, it is a three-way recursive algorithm. Let us see the following code:

```python
import turtle
def draw_triangle(points, color, my_turtle):
    my_turtle.fillcolor(color)
    my_turtle.up()
    my_turtle.goto(points[0][0],points[0][1])
    my_turtle.down()
    my_turtle.begin_fill()
    my_turtle.goto(points[1][0], points[1][1])
    my_turtle.goto(points[2][0], points[2][1])
    my_turtle.goto(points[0][0], points[0][1])
    my_turtle.end_fill()

def get_mid(p1, p2):
    return ((p1[0] + p2[0]) / 2, (p1[1] + p2[1]) / 2)

def sierpinski(points, degree, my_turtle):
    color_map = ['blue', 'red', 'green', 'white', 'yellow', 'violet', 'orange']
    draw_triangle(points, color_map[degree], my_turtle)
    if degree > 0:
        sierpinski([points[0], get_mid(points[0], points[1]), get_mid(points
[0], points[2])], degree-1, my_turtle)
        sierpinski([points[1], get_mid(points[0], points[1]), get_mid(points
[1], points[2])], degree-1, my_turtle)
        sierpinski([points[2], get_mid(points[2], points[1]), get_mid(points
[0], points[2])], degree-1, my_turtle)
```

https://doi.org/10.1515/9783110689488-011

```
def main():
    my_turtle = turtle.Turtle()
    my_win = turtle.Screen()
    my_points = [[-100, -50], [0, 100], [100, -50]]
    sierpinski(my_points, 3, my_turtle)
    my_win.exitonclick()

main()
```

Figure 11.1: The Sierpinski triangle.

In the given code we are going to use recursion. Recursion is the technique of problem-solving and calling function. In recursion, a function is called by itself again and again until a given condition is true. Here we break a problem down into smaller and smaller subproblems until we get to a problem small enough that it can be solved trivially. Let us see one more example with graphics turtle:

```
import turtle
my_turtle = turtle.Turtle()
my_win = turtle.Screen()
def draw_spiral(my_turtle, line_len):
    if line_len > 0:
        my_turtle.forward(line_len)
        my_turtle.right(90)
        draw_spiral(my_turtle, line_len - 5)

draw_spiral(my_turtle, 200)
my_win.exitonclick()
```

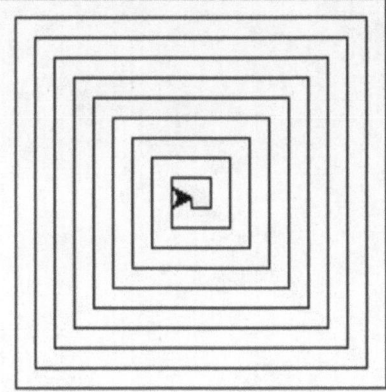

Figure 11.2: Turtle with recursion.

Fractal tree comes from a branch of mathematics and has much in common with recursion. The definition of a fractal is that, when we look at it, the fractal has the same basic concept no matter how much we magnify it. Let us see in the given code:

```python
import turtle
def tree(branch_len, t):
    if branch_len > 5:
        t.forward(branch_len)
        t.right(20)
        tree(branch_len - 15, t)
        t.left(40)
        tree(branch_len - 15, t)
        t.right(20)
        t.backward(branch_len)
def main():
    t = turtle.Turtle()
    my_win = turtle.Screen()
    t.left(90)
    t.up()
    t.backward(100)
    t.down()
    t.color("green")
    tree(75, t)
    my_win.exitonclick()
main()
```

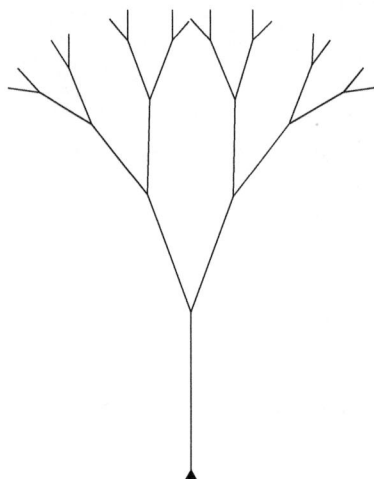

Figure 11.3: Fractal tree with turtle.

11.2 Interacting with another framework

Function *kivy.support.install_twisted_reactor* is used to install a twisted reactor that will run inside the Kivy event loop, passing argument to this function on thread selected reactor interleave function. Let us see the code where the server app has a simple twisted server running and logs messages if any. The client app can send a message to the server and will print its message and response it got. We will write code into two different modules, server.py and client.py. Let us see first server-side code:

```
from kivy.support import install_twisted_reactor
install_twisted_reactor()
from twisted.internet import reactor
from twisted.internet import protocol
class EchoServer(protocol.Protocol):
    def dataReceived(self, data):
        response = self.factory.app.handle_message(data)
        if response:
            self.transport.write(response)


class EchoServerFactory(protocol.Factory):
    protocol = EchoServer
    def __init__(self, app):
        self.app = app
----------------------------Twisted Server Code--------------------------

from kivy.app import App
from kivy.uix.label import Label
```

```python
class TwistedServerApp(App):
    label = None
    def build(self):
        self.label = Label(text="server started\n")
        reactor.listenTCP(8000, EchoServerFactory(self))
        return self.label
    def handle_message(self, msg):
        msg = msg.decode('utf-8')
        self.label.text = "received: {}\n".format(msg)
        if msg == "ping":
            msg = "Pong"
        if msg == "plop":
            msg = "Kivy Rocks!!!"
        self.label.text += "responded: {}\n".format(msg)
        return msg.encode('utf-8')
if __name__ == '__main__':
    TwistedServerApp().run()
```

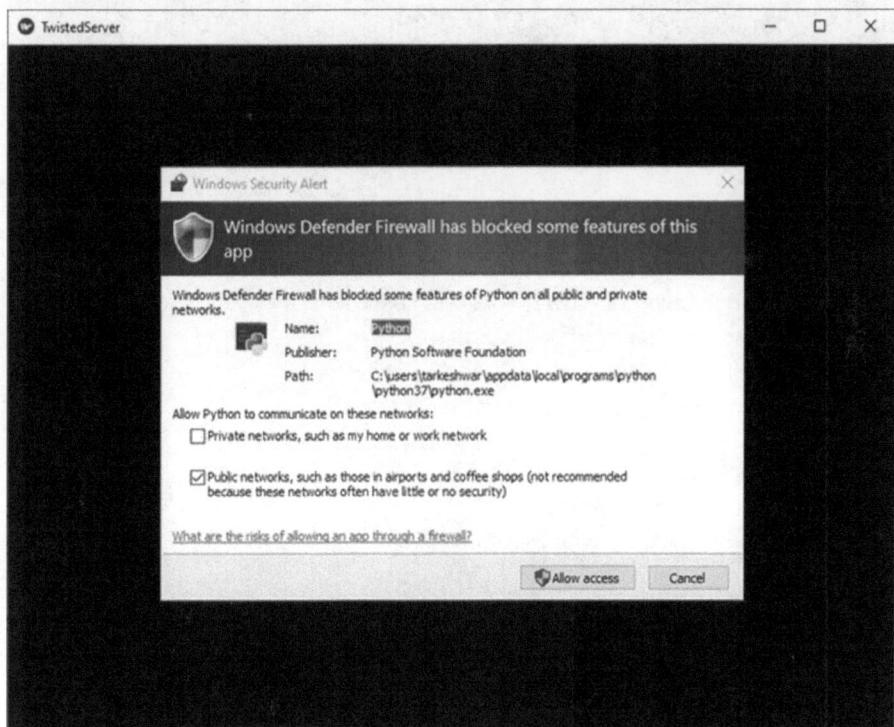

Figure 11.4: Permission popup menu to start server.

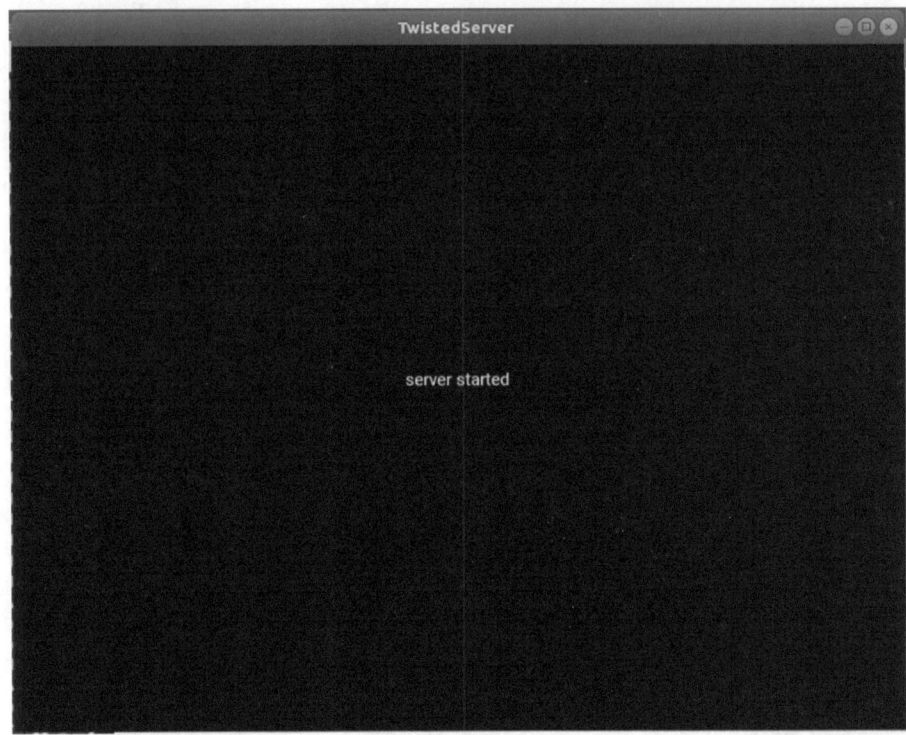

Figure 11.5: Server started.

Client.py code begins from here:

```python
from kivy.support import install_twisted_reactor
install_twisted_reactor()
# A Simple Client that send messages to the Echo Server
from twisted.internet import reactor, protocol

class EchoClient(protocol.Protocol):
    def connectionMade(self):
        self.factory.app.on_connection(self.transport)

    def dataReceived(self, data):
        self.factory.app.print_message(data.decode('utf-8'))

class EchoClientFactory(protocol.ClientFactory):
    protocol = EchoClient
    def __init__(self, app):
        self.app = app
```

```python
    def startedConnecting(self, connector):
        self.app.print_message('Started to connect.')

    def clientConnectionLost(self, connector, reason):
        self.app.print_message('Lost connection.')

    def clientConnectionFailed(self, connector, reason):
        self.app.print_message('Connection failed.')
```

---------------------------Twisted client code----------------------------

```python
from kivy.app import App
from kivy.uix.label import Label
from kivy.uix.textinput import TextInput
from kivy.uix.boxlayout import BoxLayout

# A simple kivy App, with a textbox to enter messages, and
# a large label to display all the messages received from
# the server
class TwistedClientApp(App):
    connection = None
    textbox = None
    label = None

    def build(self):
        root = self.setup_gui()
        self.connect_to_server()
        return root

    def setup_gui(self):
        self.textbox = TextInput(size_hint_y=.1, multiline=False)
        self.textbox.bind(on_text_validate=self.send_message)
        self.label = Label(text='connecting. . .\n')
        layout = BoxLayout(orientation='vertical')
        layout.add_widget(self.label)
        layout.add_widget(self.textbox)
        return layout

    def connect_to_server(self):
        reactor.connectTCP('localhost', 8000, EchoClientFactory(self))

    def on_connection(self, connection):
        self.print_message("Connected successfully!")
        self.connection = connection
```

```python
    def send_message(self, *args):
        msg = self.textbox.text
        if msg and self.connection:
            self.connection.write(msg.encode('utf-8'))
            self.textbox.text = ""

    def print_message(self, msg):
        self.label.text += "{}\n".format(msg)

if __name__ == '__main__':
    TwistedClientApp().run()
```

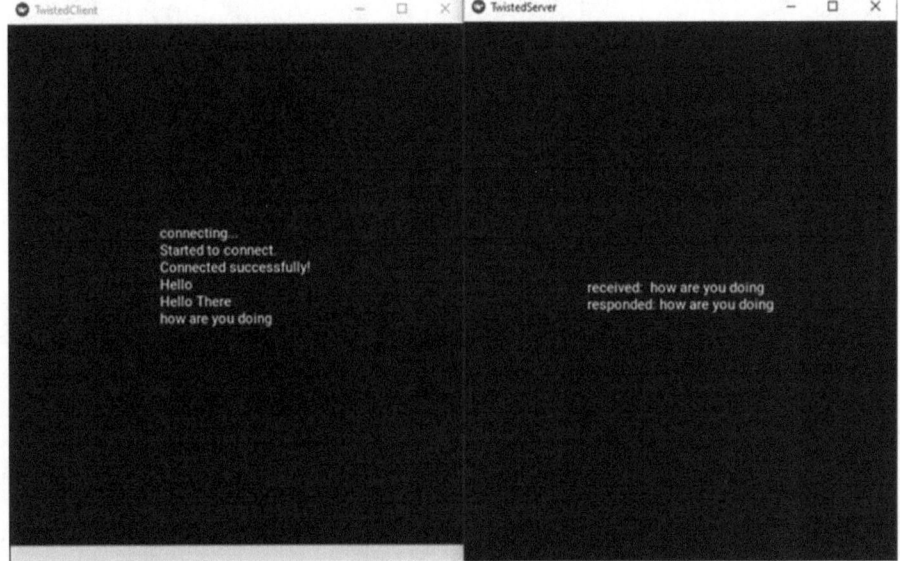

Figure 11.6: Server connection with client.

11.3 Modules

The modules are small pieces of codes that help us to write our code easily. These are Python code written files, known as modules. These are just like an add-on, plugins in a web browser, and any plugin can be added/removed at any moment without affecting the existing one. Some of the classes and modules are needed to start Kivy applications. They have to be loaded during Kivy application start from the Kivy path (PATH_TO_KIVY/kivy/modules) and user path (HOME/.kivy/mods), and a configuration file manages this. Some of the important modules are as follows:

1) **joycursor** – It is a joystick navigator to know the current location in our app.
2) **webdebugger** – It is like a testing tool to test an app using the web browser.
3) **touchring** – Its task is to draw a circle around the touched area.
4) **monitor** – It adds a red top bar that indicates the FPS and a small graph indicating input activity.
5) **showborder** – It is responsible to display a border around the widget.
6) **inspector** – It is like an examining tool to test hierarchy and widget properties.
7) **screen** – It is used to examine the screen resolution and density of various screens.
8) **recorder** – It is a sequence of events like record and playback.
9) **keybinding** – Some keys need to be bound to perform some special tasks such as screenshot.

To activate these modules, we use configuration file (HOME/.kivy/config.ini), command line, and Config statement in Python code:

```
------------------------------Config.ini------------------------------
[modules]
showborder =
inspector =
touchring =
monitor =
keybinding =
----------------------------------------------------------------------

import kivy
kivy.require('1.0.8')
#Activate the monitor, showborder, touchring, inspector module
from kivy.config import Config

Config.set('modules', 'monitor','showborder')

or

$ python configuration.py -m webdegger #to activate web debugger from command
line
```

11.4 Network support

To communicate over network, Kivy provides *UrlRequest* to make asynchronous request on the web and gets the result when the request is complete. The content is also decoded if the Content-Type is "application/json" and the result is automatically passed to *json.loads* function:

```
from kivy.network.urlrequest import UrlRequest
def got_json(req, result):
    for key, value in result['persons'].items():
        print('{}: {}'.format(key, value))

req = UrlRequest('http://localhost:5000/persons', got_json)
```

In the UrlRequest, the first argument is mandatory and the second one is optional. By default, it gets the HTTP method, and as per requirement, the method can be POST, DELETE, PUT, and so on as given in the syntax class:

```
class kivy.network.urlrequest.UrlRequest(url, on_success=None, on_redirect=None,
on_failure=None, on_error=None, on_progress=None, req_body=None, req_headers=None,
chunk_size=8192, timeout=None, method=None, decode=True, debug=False,
file_path=None, ca_file=None, verify=True, proxy_host=None, proxy_port=None,
proxy_headers=None, user_agent=None, on_cancel=None)
```

11.5 Storage tool

Storage module is used to store and retrieve any number of key-value pair with indexed key. The access of this module is not available to use directly, and it requires future version if we are working with Python2 or less than Python3.6. Some of them are given as follows:
1) **kivy.storage.dictstore.DictStore** – It is used to store Python dictionaries.
2) **kivy.storage.jsonstore.JsonStore** – It is used to store JSON files.
3) **kivy.storage.redisstore.RedisStore** – It uses Redis database to store redis-py.

A dictionary is a collection that is unordered, mutable, and indexed. In Python, dictionaries are written with braces, and they have key-value pairs like a map. The dictionary is stored as data, unlike other data types that hold only a single value as an element. Using dictionaries data becomes more readable and optimized because of key-value pairs. Each key-value pair in dictionaries is separated by : (colon) whereas each key is separated by, (comma). Dictionary's key can be anything immutable data type like integer, float, string, and tuple. Let us see the code with DictStore:

```
from kivy.storage.dictstore import DictStore

store = DictStore('myjson.json')
# put some values
store.put('programmer', name='Tarkeshwar Barua', org='BlueCrest University')
```

```
store.put('Trainer', name='Emmanuel Teage', age=34)
# using the same index key erases all previously added key-value pairs
store.put('Network', name='Taddy', age=30) # get a value using a index key and
key
print('Tarkeshwar is', store.get('programmer')['org'])
print('Tarkeshwar is exists', store.exists('programmer'))
store.delete("programmer")
print('Tarkeshwar is exists', store.exists('programmer'))
for item in store.find(name='Emmanuel Teage'):
    print('Emmanuel Teage index key is', item[0])
    print('his key value pairs are', str(item[1]))
```

```
--------------------------------output--------------------------------
Tarkeshwar is BlueCrest University
Tarkeshwar is exists True
Tarkeshwar is exists False
Emmanuel Teage index key is Trainer
his key-value pairs are {'name': 'Emmanuel Teage', 'age': 34}
----------------------------------------------------------------------
```

JSON stands for JavaScript Object Notation, it is used to store information in an organized, and easy-to-easy access manner. It is more readable and like Directory in Python, but dictionaries can be used only in Python while JSON is cross-platform. It is widely used to transmit data between various devices. Let us see in the code how to store JSON in kivy Object:

```
from kivy.storage.jsonstore import JsonStore
store = JsonStore('myjson.json')
# put some values
store.put('programmer', name='Tarkeshwar Barua', org='BlueCrest University')
store.put('Trainer', name='Emmanuel Teage', age=34)
# using the same index key erases all previously added key-value pairs
store.put('Network', name='Taddy', age=30)
# get a value using a index key and key
print('Tarkeshwar is', store.get('programmer')['org'])
print('Tarkeshwar is exists', store.exists('programmer'))
store.delete("programmer")
print('Tarkeshwar is exists', store.exists('programmer'))
for item in store.find(name='Emmanuel Teage'):
    print('Emmanuel Teage index key is', item[0])
    print('his key value pairs are', str(item[1]))
```

```
--------------------------------output--------------------------------
Tarkeshwar is BlueCrest University
Tarkeshwar is exists True
Tarkeshwar is exists False
Emmanuel Teage index key is Trainer
his key value pairs are {'name': 'Emmanuel Teage', 'age': 34}
----------------------------------------------------------------------
```

Redis is an open-source, BSD-licensed, advanced key-value store. It is often referred to as a data structure server since the keys can contain strings, hashes, lists, sets, and sorted sets. Redis is written in C language. To store implementation using Redis, we must have installed redis-py. Let us have a look at the following code:

```python
import redis
pool=redis.ConnectionPool(host='192.168.255.148',port=6379,db=0)
r = redis.Redis(connection_pool=pool)
r.set('foo', 'bar')
print(r.get("foo"))
r.mset({"FirstName": "Tarkeshwar", "LastName": "Barua"})
r.set("Country", "INDIA")
print("FirstName : ", r.get("FirstName"))
print("LastName : ", r.get("LastName"))
print("foo Key : ", r.get("foo"))
print("Country : ", r.get("Country"))

--------------------------------output--------------------------------
b'bar'
FirstName : b'Tarkeshwar'
LastName : b'Barua'
foo Key : b'bar'
Country : b'INDIA'
----------------------------------------------------------------------
```

11.6 Rotate clock and rotate speed

The Clock Object allows us to schedule a function call in the future, once or repeatably at specified intervals. We can get the time elapsed between the schedule and the calling of the callback via dt argument. The callback is a weak reference. We are responsible to keep the reference to our original object to execute our callback. The clock will execute all the callbacks with a timeout of −1 before the next frame even if we add a new callback with −1 from a running back. The given clock has its iteration limit for callback (default is 10). If we schedule a callback that schedules a callback as more then 10 then it will leave the loop and send a warning to the console, and then continue after the next frame:

```python
from kivy.app import App
from kivy.lang import Builder

kv = '''
FloatLayout:

Button:
  text: 'Python Kivy'
  font_size:35
  background_color:1,0,1,1
  color:0,1,1,1
  size_hint: 1,1
  pos_hint: {'center_x': .5, 'center_y': .5}
  canvas.before:
    PushMatrix
    Rotate:
      angle: 330
      origin: self.center
  canvas.after:
    PopMatrix
'''

class RotationApp(App):
    def build(self):
        return Builder.load_string(kv)

RotationApp().run()
```

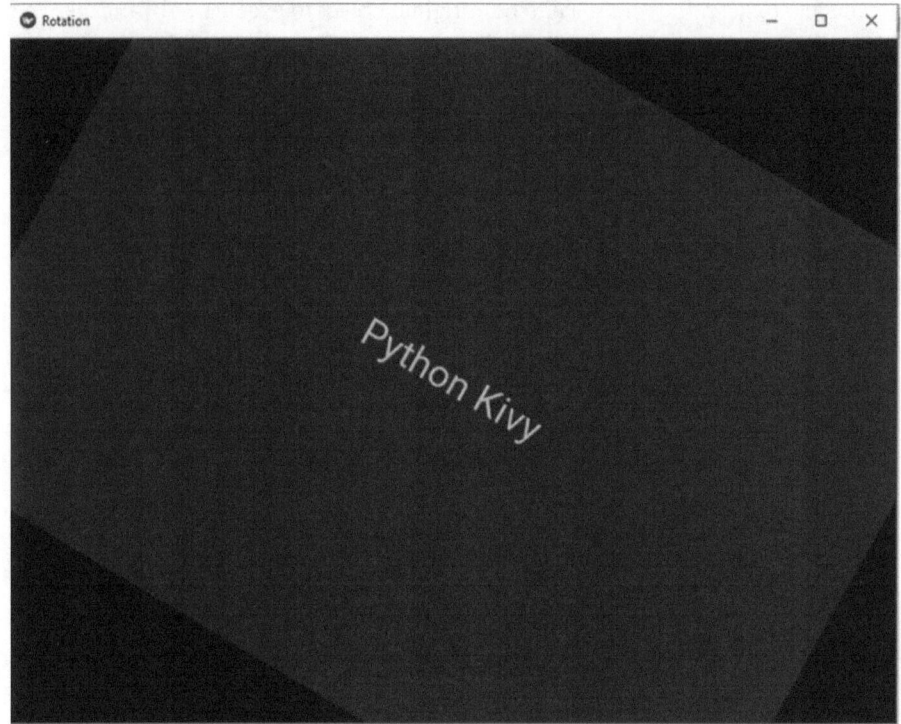

Figure 11.7: Clocking and rotating object.

11.7 Triggered events

A triggered event is a way to defer a callback. It functions exactly like *schedule_once()* and *schedule_interval()* except that it does not immediately schedule the callback. It ensures that we can call the event multiple times, but it would not schedule more than once:

```python
from kivy.app import App
from kivy.clock import Clock
from kivy.uix.label import Label
class ClockDemo(App):
    count = 0
    def build(self):
        self.myLabel = Label(text='Count down Start. . .')
        self.myLabel.color=1,1,0,1
        self.myLabel.font_size=100
```

```
        Clock.schedule_interval(self.Callback_Clock, 2)
        return self.myLabel

    def Callback_Clock(self, dt):
        self.count = self.count + 1
        self.myLabel.text = str(self.count)

if __name__ == '__main__':
    ClockDemo().run()
```

Figure 11.8: Event trigger.

And unscheduling can be done by the following code:

```
def my_callback(dt):
    pass

# call my_callback every 0.5 seconds
event = Clock.schedule_interval(my_callback, 0.5)
```

```
# call my_callback in 5 seconds
event2 = Clock.schedule_once(my_callback, 5)

event_trig = Clock.create_trigger(my_callback, 5)
event_trig()

# unschedule using cancel
event.cancel()

# unschedule using Clock.unschedule
Clock.unschedule(event2)

# unschedule using Clock.unschedule with the callback
# NOT RECOMMENDED
Clock.unschedule(my_callback)
```

All the scheduling and canceling methods are fully thread-safe and can be safely used from external threads. Kivy clock attempts to execute any schedule callback.

11.8 Animations

Animation and AnimationTransition are used to animate widget properties. We must specify at least a property name and target value. This animation will last for 1 sec unless duration is specified when *anim.start()* is called; the widget will move smoothly from the current position to (100, 100) location. We can animate multiple properties and use built-in or custom transition functions using transactions. To animate a widget, we can use the following code:

```
from kivy.animation import Animation
from kivy.app import App
from kivy.uix.button import Button
from kivy.uix.label import Label
from kivy.uix.pagelayout import PageLayout
from kivy.uix.widget import Widget

class MainApp(App):
    def printme(self,a):
        print("Yes Button has been Clicked",a)

    def printed(self,b):
        print("button released",b.text)

    def statetest(self,x,y):
        print("State Geting Changed")
```

```
    def build(self):
        btn=Button(text="Click Me", font_size=30, color=(1, 0, 0, 1), size_-
hint=(0.2,0.2), background_color=(0, 1, 0, 1))
        lbl = Label(text="Good Morning", font_size=40, color=(0, 1, 0, 1),
#size_hint=(0.2, 0.2), #background_color=(0, 0, 1, 1))

        btn.bind(on_press=self.printme, on_release=self.printed, state=self.
statetest)
        layout=PageLayout()
        anim = Animation(x=100, y=100, t="in_quad", duration=2)
        anim.start(btn)
        anim.start(lbl)
        layout.add_widget(btn)
        layout.add_widget(lbl)
        return layout
if __name__=='__main__':
    MainApp().run()
```

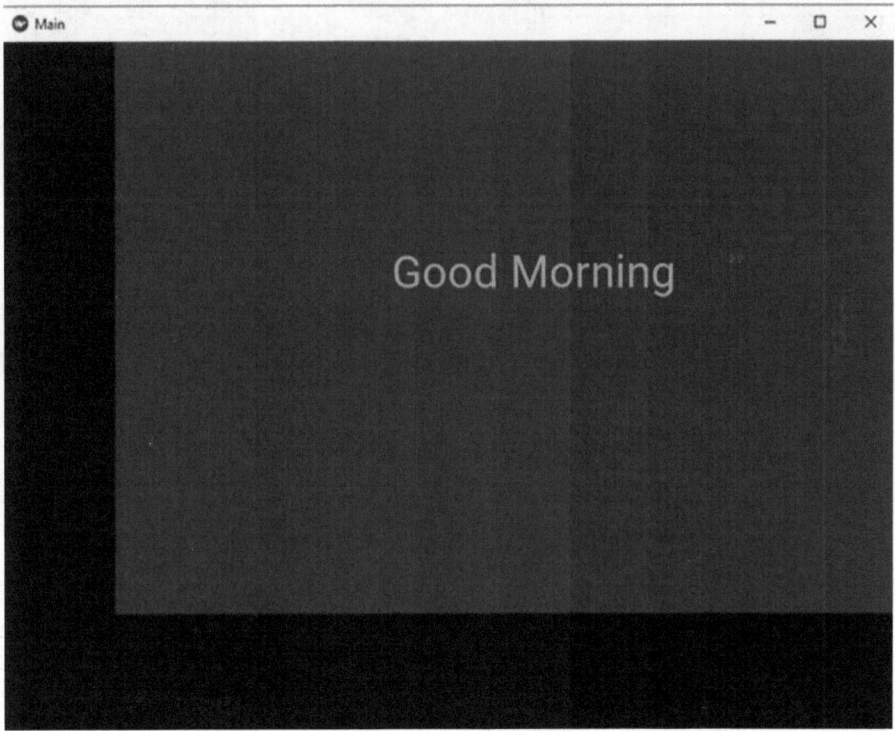

Figure 11.9: Animation with trigger.

11.9 Audio

In Kivy we can load and play an audio file. Class SoundLoader.load() will be the best sound provider for that file type depends upon the different sound file types. Event can also be handled over sound such as event loop:

```
#file name audio.kv
<RootScreen>
SoundBoard:
<SoundBoard>
name: 'soundboard'
BoxLayout:
  Button:
    text: 'click here to play music'
    font_size:70
    background_color:1,0,0,1
    color:1,1,0,1
    on_press: root.scotch()
---------------------------MainApp-----------------------------------
from kivy.app import App
from kivy.uix.label import Label
from kivy.uix.boxlayout import BoxLayout
from kivy.properties import ListProperty, ObjectProperty, NumericProperty
from kivy.uix.screenmanager import ScreenManager, Screen
from kivy.core.audio import SoundLoader


class SoundBoard(Screen):
    def scotch(self):
        sound = SoundLoader.load('song.wav')
    if sound:
        sound.play()
        sound.seek(0.00)
        print("Sound found at %s" % sound.source)
        print("Sound is %.3f seconds" % sound.length)
        print("Sound status %s" % sound.status)


class RootScreen(ScreenManager):
    pass


class audio(App):
    def build(self):
        return SoundBoard()
```

```
if __name__ == "__main__":
    audio().run()
```

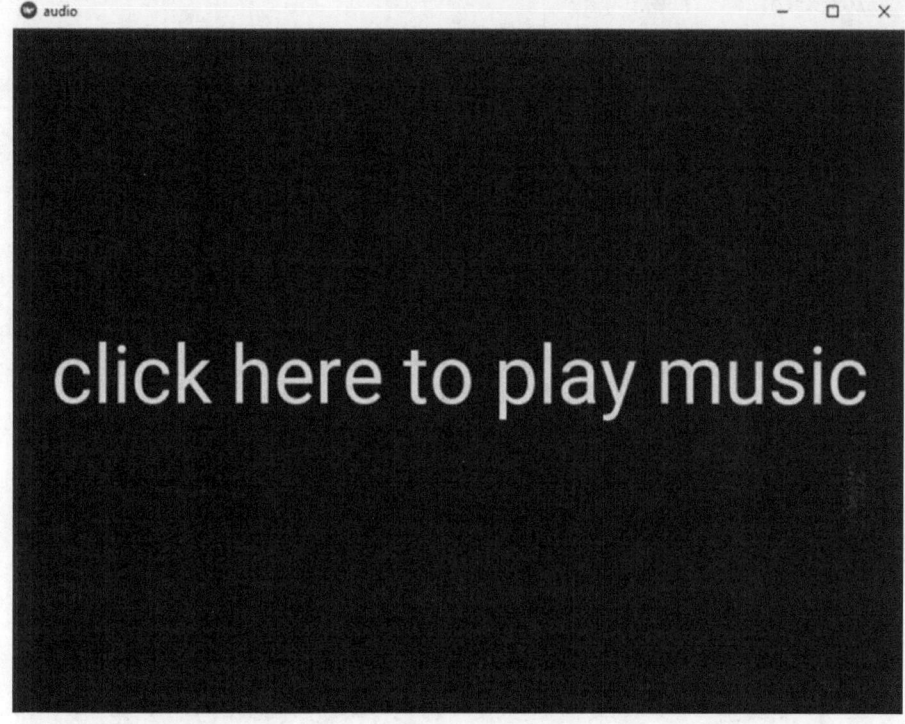

Figure 11.10: Playing audio.

11.10 Video

In Kivy Video widget is available to display video files and streams. We can control the entire video such as play, pause, volume, and position. Depending on our video core provider, platform and plugins, we will be able to play various formats. The widget can't be customized much because of the complex assembly of numerous base widgets. Pygame video provider allows only MPEG1 on Linux and OSX. GStreamer is moving versatile and can read many video containers and codes such as MKV, OGV, AVI, MOV, and FLV. Video loading is asynchronous meaning many properties may not available until the video is loaded. Let us see the given code:

```
from kivy.app import App
from kivy.uix.videoplayer import VideoPlayer
```

```python
class MyVideoApp(App):
    def build(self):
        self.player = VideoPlayer(source='video.mkv', state='play', options=
{'allow_stretch': True})
        return (self.player)
if __name__ == '__main__':
    MyVideoApp().run()
```

Figure 11.11: Playing video https://kivy.org/doc/stable/api-kivy.uix.videoplayer.html.

Let us see another example to play video with kv language:

```
#:kivy 1.9.0
<ScreenOne>:
    name: "screen1"
    ScrollView:
        pos_hint: {'bottom':1}
        do_scroll_y: True
        GridLayout:
            name: "screen1"
            cols: 1
            padding: 0
```

```
                    spacing: 0
                    #size_hint: None, None
                    size_hint_y: None
                    height: self.minimum_height
                    Button:
                        text: "1"
                        background_color: (0.502, 0, 0.502, 1) if self.state
== 'normal' else (0.502, 0, 0.502, .75)
                        on_release: root.onNextScreen(self, 'small')
                    Button:
                        text: "2"
                        background_color: (0, 0, 1, 1) if self.state == 'normal'
else (0, 0, 1, .75)
                        on_release: root.onNextScreen(self, 'sample1')
                    Button:
                        text: "3"
                        background_color: (0.251, 0.878, 0.816, 1) if self.
state == 'normal' else (0.251, 0.878, 0.816, .75)
                        on_release: root.onNextScreen(self, 'sample2')
                    Button:
                        text: "4"
                        background_color: (0, 0.502, 0, 1) if self.state == 'nor-
mal' else (0, 0.502, 0, .75)
                        on_release: root.onNextScreen(self, 'sample3')
                    Button:
                        text: "5"
                        background_color: (1, 1, 0, 1) if self.state == 'normal'
                        else (1, 1, 0, .75)
                        on_release: root.onNextScreen(self, 'sample4')
                    Button:
                        text: "6"
                        background_color: (1, 0.843, 0, 1) if self.state == 'nor-
mal' else (1, 0.843, 0, .75)
                        on_release: root.onNextScreen(self, 'sample5')
                    Button:
                        text: "7"
                        background_color: (1, 0.647, 0, 1) if self.state == 'nor-
mal' else (1, 0.647, 0, .75)
                        on_release: root.onNextScreen(self, 'sample6')
                    Button:
                        text: "8"
```

```
                            background_color: (1, 0.549, 0, 1) if self.state ==
'normal' else (1, 0.549, 0, .75)
                            on_release: root.onNextScreen(self, 'sample7')
                    Button:
                            text: "9"
                            background_color: (1, 0, 0, 1) if self.state == 'nor-
mal' else

                            (1, 0, 0, .75)
                            on_release: root.onNextScreen(self, 'sample8')
<ScreenTwo>:
        GridLayout:
                name: "ScreenGrid"
                pos_hint: {'top': 1}
                cols:1
                rows:2
                ActionBar:
                        pos_hint: {'top': 1}
                        height:'100sp'
                        ActionView:
                                ActionPrevious:
                                        with_previous: True
                                        on_release: root.onBackBtn()
<Manager>:
        id: screen_manager
        screen_one: screen_one
        screen_two: screen_two
        ScreenOne:
                id: screen_one
                name: 'screen1'
                manager: screen_manager
        ScreenTwo:
                id: screen_two
                name: 'screen2'
                manager: screen_manager
<VideoPlayer>:
        options: {'eos': 'loop'}
        allow_stretch:True
<Button>:
        height: '100sp'
        size_hint_y: None
```

```
-------------------------------MainApp-----------------------------
import kivy
kivy.require('1.9.0')
from kivy.app import App
from kivy.uix.screenmanager import ScreenManager, Screen, NoTransition
from kivy.properties import ObjectProperty
from kivy.uix.videoplayer import VideoPlayer
from kivy.uix.actionbar import ActionBar
from kivy.uix.button import Button

class ScreenTwo(Screen):
    def test_on_enter(self, vidname):
        #self.add_widget(Button(text="Back"))
        self.vid = VideoPlayer(source=vidname, state='play', options={'al-
low_stretch':False, 'eos': 'loop'})
        self.add_widget(self.vid)

    def on_leave(self):
        pass

    def onBackBtn(self):
        self.vid.state = 'stop'
        self.remove_widget(self.vid)
        self.manager.current = self.manager.list_of_prev_screens.pop()

class ScreenOne(Screen):
    def onNextScreen(self, btn, fileName):
        self.manager.list_of_prev_screens.append(btn.parent.name)
        self.manager.current = 'screen2'
        self.manager.screen_two.test_on_enter('2.mp4')

class Manager(ScreenManager):
    transition = NoTransition()
    screen_one = ObjectProperty(None)
    screen_two = ObjectProperty(None)
    screen_three = ObjectProperty(None)

    def __init__(self, *args, **kwargs):
        super(Manager, self).__init__(*args, **kwargs)
        # list to keep track of screens we were in
        self.list_of_prev_screens = []
```

```
class Screen_testApp(App):
    def build(self):
        return Manager()

if __name__ == "__main__":
    Screen_testApp().run()
```

Annotation

Annotations are used to display text at a time and duration. Annotation files are based on JSON. The player loads it automatically like .kv file. Annotation files are saved with the extension name ".jsa." How to do basic file operations like read–write, open file using built and standard libraries, let us see the syntax:

```
[
    {"start": 0, "duration": 10,
    "text": "My First annotation"},
    {"start": 1, "duration": 20,
    "bgcolor": [0.2, 0.1, 0.3, 0.4],
    "text": "background color changible"}
]
```

Summary

-   Graphics make our app interactive and attractive. Python provides graphics module to do so.
-   The recursion is a way of resolving problem where we break a problem down into smaller and smaller subproblems until we get a problem small enough and can be solved trivially.
-   Fractal tree comes from mathematics and has much in common with recursion.
-   Function kivy.support.install_twisted_reactor is used to install a twisted reactor that will run inside the Kivy event loop, passing argument to this function on thread selected reactor interleave function.
-   The modules are small pieces of codes that we write in Python file (.py).
-   These have to be loaded during Kivy application. Starting from the Kivy path (PATH_TO_KIVY/kivy/modules) and user path (HOME/.kivy/mods), this is managed by a configuration file.
-   The Kivy needs some modules to be loaded at the time of execution. Some of them are joycursor, webdebugger, touchring, monitor, showborder, inspector, screen, recorder, and keybinding.

– These modules are associated with file (HOME/.kivy/config.ini) and can be activated using it.
– The network communication can be established between other devices with the help of *UrlRequest*. In the *UrlRequest* function one first argument is mandatory. This function works asynchronously. The function json.loads decodes content into "application/json." It supports all the HTTP functions.
– The key-value pair with indexed keys data can be stored and retrieved with the storage module. It is only available with Python 3.7 and above. The storage modules are kivy.storage.dictstore.DictStore, kivy.storage.jsonstore.JsonStore, and kivy.storage.redisstore.RedisStore
– A dictionary is a collection that is unordered, mutable, and indexed. In Python, dictionaries are written with braces, and they have key-value pairs like a map. The dictionary stores data, unlike other data types that hold only a single value as an element. Using dictionaries data becomes more readable and optimized because of key-value pairs.
– JSON stands for Java Script Object Notation, it is used to store information in an organized, and easy-to-easy access manner. It is more readable and similar to Directory in Python, but dictionaries can be used only in Python while JSON is cross-platform.
– Redis is an open-source, BSD-licensed, advanced key-value store. It is often referred to as a data structure server since the keys can contain strings, hashes, lists, sets, and sorted sets. Redis is written in C language. To store implementation using Redis, we must have installed redis-py.
– The Clock Object allows us to schedule a function call in the future, once or repeatedly at specified intervals. We can get the time elapsed between the schedule and the calling of the callback via dt argument. The callback is a weak reference. We are responsible to keep a reference to our original object to execute our callback.
– A triggered event is a way to defer a callback. It functions exactly like schedule_once() and schedule_interval() except that it does not immediately schedule the callback. It ensures that we can call the event multiple times, but it would not schedule more than once.
– Animation and AnimationTransition are used to animate widget properties. We must specify at least a property name and target value. This animation will last for 1 sec unless duration is specified when anim.start() is called; the widget will move smoothly from the current position to 100, 100.
– The class SoundLoader.load() will be the best sound provider for that file type depends upon the different sound file types. Event also can be handled over sound such as event loop.
– Kivy Video widget is available to display video files and streams. We can control the entire video such as play, pause, volume, and position. Depending on

our video core provider, platform, and plugins, we will be able to play various formats.

- Annotations are used to display text at a time and duration. Annotation files are based on JSON. The player loads it automatically like .kv file. Annotation files are saved with the extension name ".jsa."

Key terms

Graphics programming requires lots of effort while we are using graphics module things become very easy. The recursion is a problem-resolving technique where we break a problem down into smaller and smaller subproblems until we get to a problem small enough that it can be solved trivially. The server app is running which logs messages if any and the client app can send message to the server and will print message and response to a client. The modules are small pieces of codes that we write in Python file (.py). They have to be loaded during Kivy application starting from the Kivy path (PATH_TO_KIVY/kivy/modules) and user path (HOME/.kivy/mods), a configuration file manages this. The Kivy needs some modules to be loaded at the time of execution. These modules are associated with file (HOME/.kivy/config.ini) and can be activated using it. The network communication can be established between other devices with the help of UrlRequest. In the UrlRequest function, one first argument is mandatory. This function works asynchronously. The function json.loads decodes content into "application/json.' It supports GET, PUT, POST, DELETE, and so on. The key-value pair with indexed keys data can be stored and retrieved with the storage module. It is only available with Python 3.7 and above. The storage modules are kivy.storage.dictstore.DictStore, kivy.storage.jsonstore.JsonStore and kivy.storage.redisstore.RedisStore. In the dictionary, each key-value pair is separated by: (colon) and each key separated by, (comma). The dictionaries key can be anything immutable data type such as int, string, float, and tuple. JSON is widely used to transmit data between devices because it is more readable and easier to consume and produce. Redis is often referred to as a data structure server since the keys can contain strings, hashes, lists, sets, and sorted sets. Redis is written in C language. Store implementation using Redis, we must have installed redis-py. The Clock Object allows us to schedule a function call in the future, once or repeatedly at specified intervals. We can get the time elapsed between the schedule and the calling of the callback via dt argument. The callback is a weak reference. We are responsible to keep a reference to our original object to execute our callback. A triggered event is a way to defer a callback. It functions exactly like schedule_once() and schedule_interval() except that it does not immediately schedule the callback. It ensures that we can call the event multiple times, but it would not schedule more than once. Animation and AnimationTransition are used to animate widget properties. We must specify at least a property name and target value.

The class SoundLoader.load() will be the best sound provider for that file type depends upon the different sound file type. Kivy Video widget is available to display video files and streams. We can control entire video such as play, pause, volume, and position. Annotations are used to display text at a time and duration. Annotation files are based on JSON. The player loads it automatically like .kv file. Annotation files are saved with the extension name ".jsa."

Review questions

1. What is the role of the graphics module in Kivy?
2. How do we handle events in Kivy?
3. What is recursion and how can we use it in our program?
4. How are modules useful in programming?
5. What is the difference between the user path and kivy path?
6. Write all the modules that are needed to be loaded at the time of execution.
7. Which file is responsible to activate modules and how would we do that?
8. How to use UrlRequest?
9. List all the storage modules in kivy.
10. What are dictionaries and how are they useful in Kivy object storage?
11. Why is JSON widely used by programmers?
12. What is the Redis server and what is the mechanism to store data at the server?
13. What is a Clock object?
14. How do we handle event triggers?
15. Write a program to design animation.

Exercise

Tick the correct option

Q.1. Which is correct format of writing JSON name/value pair _____?
    a) "name":"value"
    b) name="value"
    c) name='value'
    d) 'name:value'

Q.2. Which modules are necessary to load at Kivy application start _____?
    a) Keybinding
    b) Threading
    c) Mouse
    d) All of the above

Q.3. What is a module in Python _____?
   a) A kivy file
   b) A JSON file
   c) Graphics annotation file
   d) A Python program file

Q.4. What is fractal tree _____?
   a) A tree data structure
   b) A plant
   c) A big tree
   d) Banyan tree

Q.5. Which device does provide positional information _____?
   a) Input devices
   b) Output devices
   c) Pointing devices
   d) Both a and c

Q.6. Which of the problem cannot be resolved using recursion _____?
   a) Problem without base case
   b) Length of a string
   c) Fibonacci number up to nth position
   d) Factorial of a number

Q.7. Which file is responsible for configuration _____?
   a) HOME/.kivy/config.kv
   b) HOME/.kivy/config.cnf
   c) HOME/.kivy/config.ini
   d) HOME/.kivy/config.yml

Q.8. Which is not a storage module?
   a) kivy.storage.dictstore.DictStore
   b) kivy.storage.redisstore.RedisStore
   c) kivy.storage.jsonstore.JsonStore
   d) kivy.storage.tuplestore.TupleStore

Q.9. Which HTTP protocol is not supported by UrlRequest function _____?
   a) GET
   b) PUT
   c) POST
   d) All of the above

Q.10. What is the extension name associated with annotations _____?

    a) .kv

    b) .jsa

    c) .py

    d) .ipynb

Answers

Q.1. a) "name":"value"

Q.2. a) Keybinding

Q.3. d) a Python program file

Q.4. a) A tree data structure

Q.5. b) Output device

Q.6. a) Problem without base case

Q.7. c) HOME/.kivy/config.ini

Q.8. d) kivy.storage.tuplestore.TupleStore

Q.9. d) All of the above

Q.10. b) .jsa

Fill in the blanks

1. Recursion function is called by itself again and again until the given condition is _____.

2. _____comes from a branch of mathematics.

3. _____is a testing tool to test an app using web browser.

4. The content is also decoded if the Content-Type is _____and the result automatically is passed to jsonloads function.

5. Each key-value pair in a dictionary is separated by _____whereas each key is separated by_____.

Answers

1. true

2. Fractal tree

3. webdebugger

4. application/json

5. :(colon), (comma)

12 Packaging app for various platforms

Python is a truly wonderful language. When somebody comes up with a good idea to take about 1 minute and five lines to program something that almost does what we want. Then it takes one an hour to extend the script to 300 lines, after which it still does almost what you want.
                                                                                    ~Jack Jansen

12.1 Adding behavior with mixin class

This module implements behavior that can be mixed in with the existing base widget. The ButtonBehavior mixin class provides button behavior. The behavior can be added locally and globally. We can combine this class with widgets, such as an Image, to provide alternative buttons that preserve Kivy button behavior. Let us suppose we want all background colors should be common in every built-in widget automatically. The idea behind these classes is to encapsulate properties and events associated with the type of widgets. This module implements behaviors that can be mixed in with existing widgets. Isolating these properties and events in a mixin class allows us to define our own implementation for standard Kivy widgets that can act as drop-in replacements. This means we can restyle and redefine widgets as desired without breaking compatibility as long as they implement the behaviors correctly. They can simply replace the standard widgets. Let us have a look on the given code:

```python
from kivy.app import App
from kivy.uix.behaviors import ButtonBehavior
from kivy.uix.button import Button
from kivy.uix.image import Image

class IconButton(ButtonBehavior, Image):
    def on_press(self):
        print("on_press")

    def state_changed(*args):
        print('state changed')

class AddingBehaviorApp(App):
    def build(self):
        return IconButton()

if __name__=='__main__':
    AddingBehaviorApp().run()
```

https://doi.org/10.1515/9783110689488-012

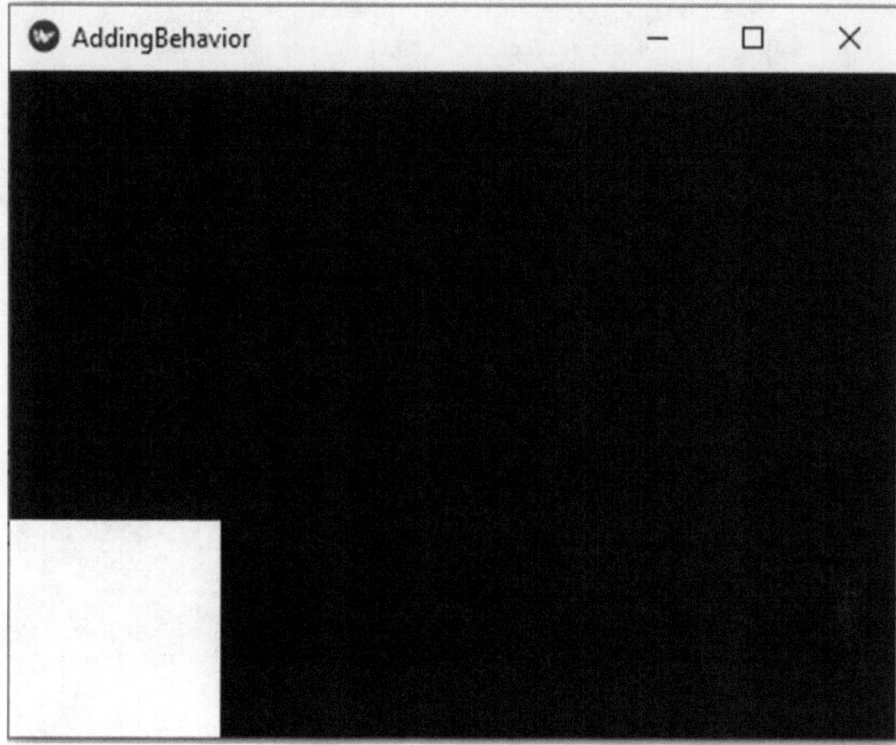

Figure 12.1: Adding behavior to components.

Let us see another example with Kv language:

```
from kivy.app import App
from kivy.lang import Builder

kv = '''
FloatLayout:
    # Define the root widget
    ScatterLayout:
        size_hint: 1.0, 0.2
        do_translation_y: False
        Label:
            size_hint: 1.0, 1
            text: 'You can Drag me'
            canvas.before:
                    Color:
                        rgb: 0, .6, .6
```

```
            Rectangle:
                pos: self.pos
        size: self.size
'''

class RectangleApp(App):
    def build(self):
        object = Builder.load_string(kv)
        return object

if __name__ == '__main__':
    RectangleApp().run()
```

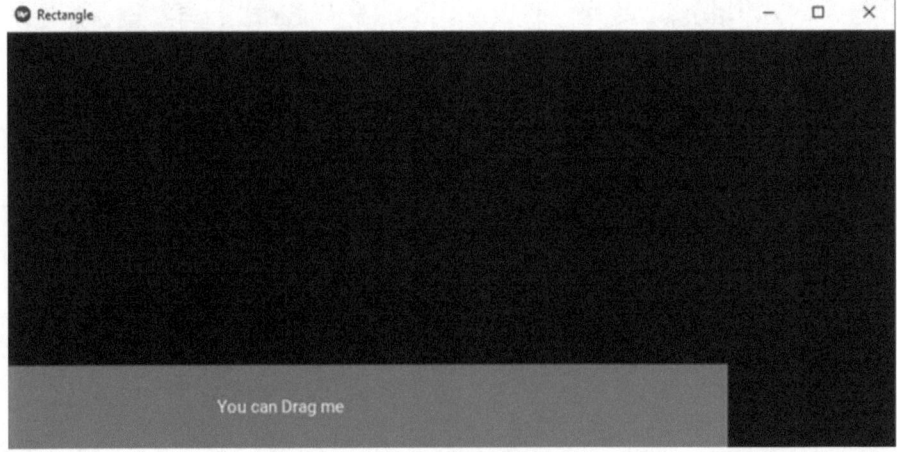

Figure 12.2: Adding behavior to components with kv language.

12.1.1 Packaging application

There are various ways to package Kivy application that depends on factors such as the requirement, location, platform, and the plan to execute the application. Packaging application for distribution is based on the platform, means generating platform should be the same for which we are generating. Packaging could be either 32 bit or 64 bit, which depends on the Python version and platforms such as Windows, Linux, Android, and iOS. **PyInstaller** is one of the tools that is used for packing. We can install pyinstaller using pip with the following procedure:

```
$pip install pyinstaller
$pyinstaller myapp.py        #to generate multiple file application
$pyinstaller --onefile myapp.py     #to generate one executable file application
```

```
$pyinstaller hello.py --icon=path/to/icon.ico    #to set specific icon to our
application
$pyinstaller hello.py --noconsole      #to generate app only for GUI
$pyinstaller hello.spec hello.py       #to generate app with given specifica-
tion file
```

Other changes can be done with *.spec* file to generate the required output file. In this example, we use pyinstaller 3.5. The executable file generation process takes some time. After completion of this process, the executable file can be located in the *dist* folder. If possible, always try to use the latest version of Kivy, which provides new features and easy-to-test application.

12.1.2 Testing application with android device

Here is a simple "Hello World" program to be tested on android platform. We require to connect our android device (Physical or emulator). In case of physical device, we need to make sure that we have enabled "USB debugging connected" and "USB connected." If it is not, then we can go to Settings→ Application →Development. We are assuming *Python4Android* is at class path to the *Python for Android* installation, and the current directory is called "testapp," the given script after the application code will build a debug APK and install it on the phone as given in Figure 12.3.

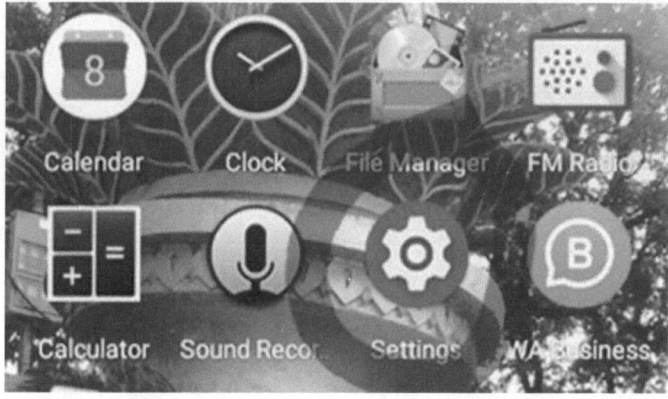

Figure 12.3: Settings on home screen.

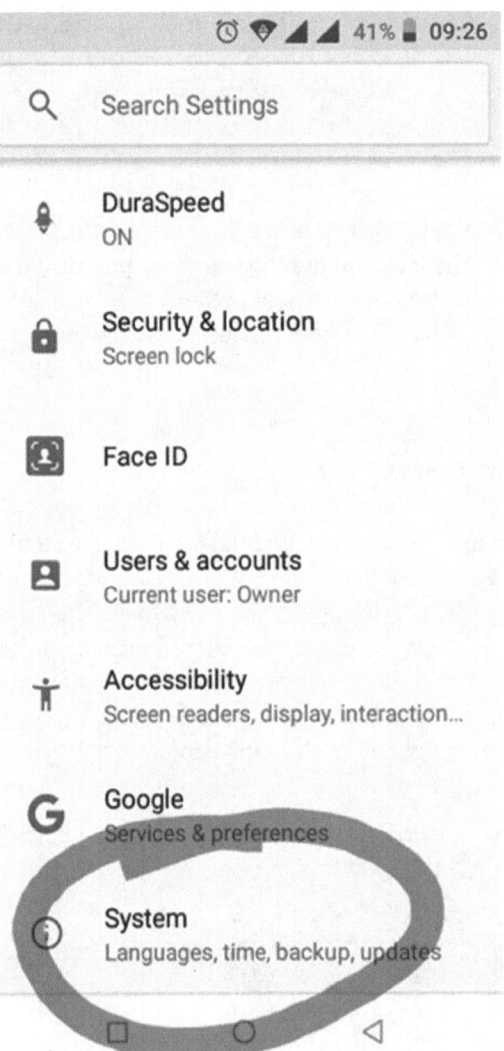

Figure 12.4: Systems in settings menu.

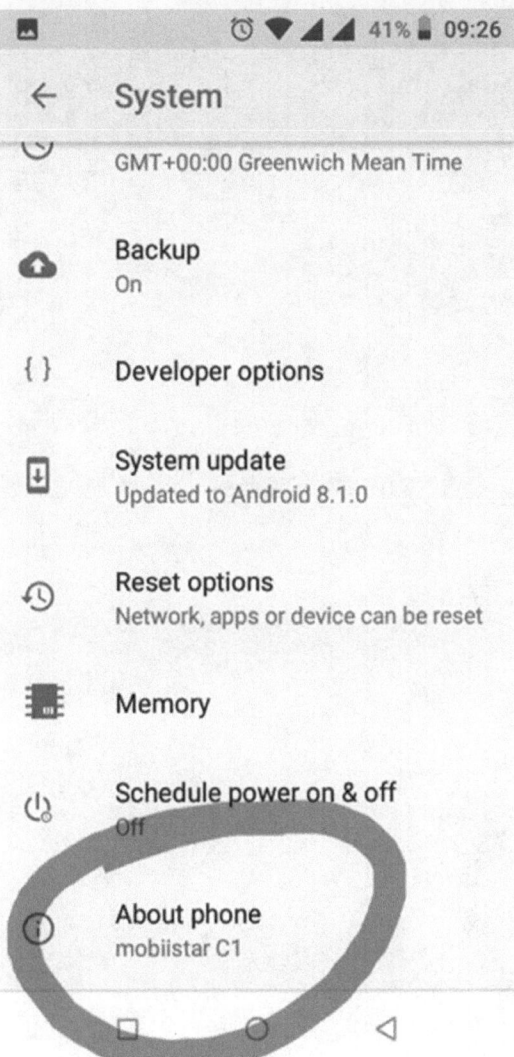

Figure 12.5: About phone in systems menu.

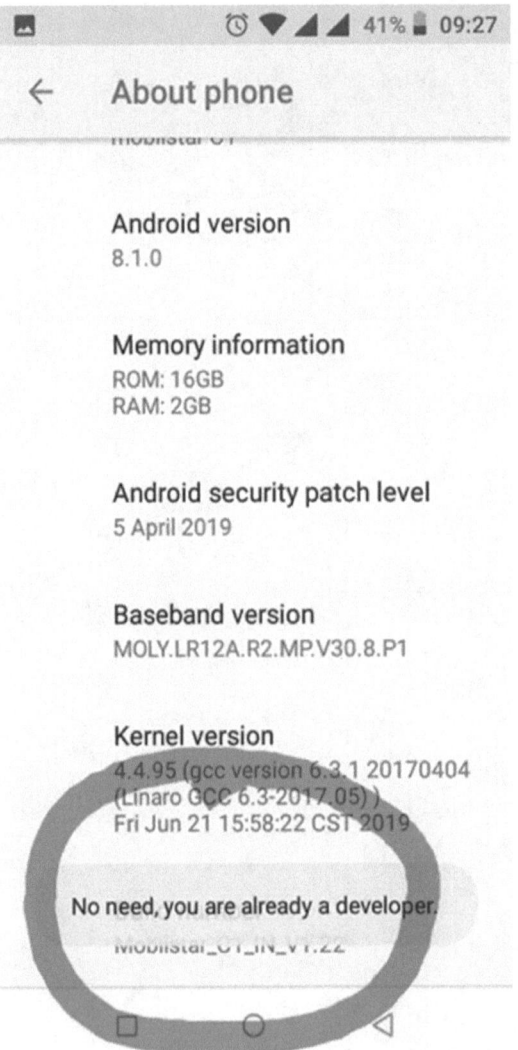

Figure 12.6: Build number in about phone menu.

We need to click (press) here five times and then it will display a toast message on the screen "You are developer now." We are already, it is enabled then it will display "you are already developer." This is the only one-time job on our phone. Now our phone is ready to test android apps with the computer. While we are developing an app, it may display a dialog box "Allow to Run Application," and then we just allow it and proceed to test our app. We need to wait until our app displays at the device screen. The testing device can be any physical and virtual devices.

```
from kivy.app import App
from kivy.uix.button import Button

class TestApp(App):
    def build(self):
        return Button(text='Hello World')
if __name__ in ('__main__', '__android__'):
    TestApp().run()
```

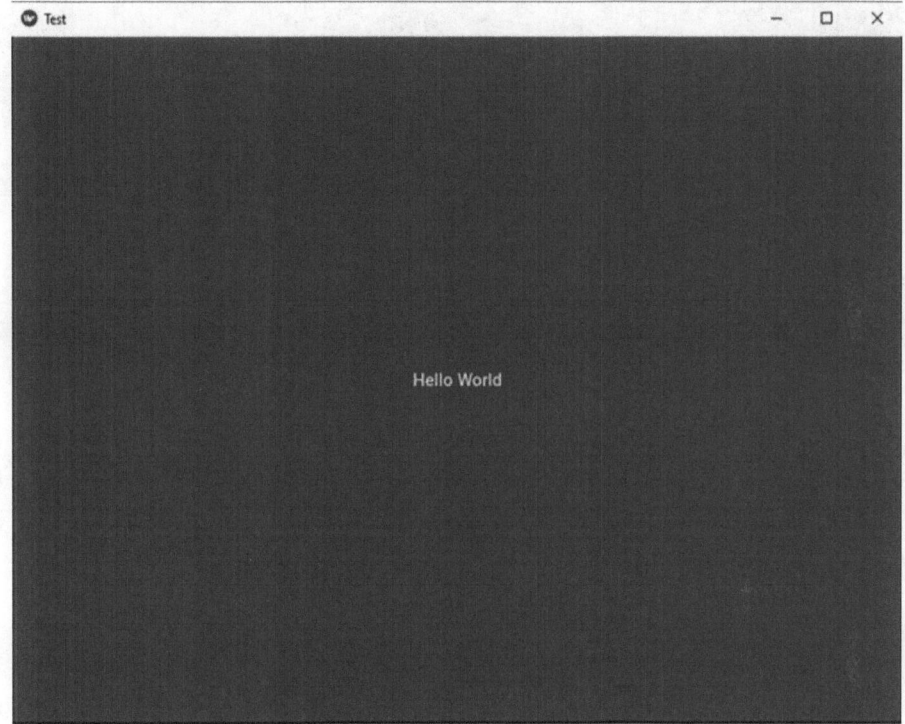

Figure 12.7: Sample app for packaging.

The APK will be built in "$Python4Android/dist/default/bin," and it appears to be necessary to remove the previous build in order to get a new version. The screen height and width can be changed by config file "$HOME/.kivy". The default orientation is landscape, but it can be changed by using the command "--orientation portrait" and the icon can be replaced with "--icon $currdir/test_icon.png."

```
currdir='pwd'

cd $Python4Android/dist/default

rm bin/testapp*
```

```
./build.py --package org.test.myapp --name myapp --version 1.0 --dir $currdir
debug
adb install -r bin/testapp-1.0-debug.apk
```

Figure 12.8: Android sample application snapshot.

As shown in Figure 12.8, Kivy app has been installed on an Android device. It may take some time to generate APK and installation of our app. Before releasing APK over the Play Store we need to sign it first. We can sign it by using the given command:

```
/build.py --package com.test.testapp --name testapp --version 1.0 --dir $curr-
dir release
jarsigner -verbose -keystore your_key_store testapp-1.0-release-unsigned.apk
your_key
jarsigner -verify testapp-1.0-release-unsigned.apk
zipalign -v 4 testapp-1.0-release-unsigned.apk testapp.apk
adb install -r testapp.apk
```

In the code, first command builds a release version. The second command signs the app and the third one verifies the signing. The fourth one aligns bytes in the app's files. The last command installs the app on the device.

12.1.3 Packaging for android

Android packaging is important because of its popularity in the market. **Python-for-android** tool is a popular packaging tool for Android. It is only supported by Linux. Python-for-android supports both Python2 and Python3 versions. We can package our fully optimized app for multiple architectures. It can be installed with the given command:

```
$pip install python-for-android
or
$pip install git+https://github.com/kivy/python-for-android.git
```

After setting SDK and NDK for android, we can build APK with the given command:

```
$p4a apk --requirements=kivy --private /home/tarkeshwar/devel/planewave_-
frozen/ - package=net.inclem.planewavessdl2 --name="planewavessdl2" --
version=0.5 --bootstrap=sdl2
```

But **Buildozer** is one of the tools that automates this entire process. The easiest way to android packaging is **Buildozer** to make full APK. We can also run our app without a compilation step with Kivy Launcher. These apps can be released over all the available platforms as a signed app.

Buildozer is a tool that automates the entire process. In this example, we are using buildozer version Buildozer 0.40.dev0. It downloads and sets up all the perquisites for Python-for-android, including the Android SDK and NDK, then builds an APK that can be automatically pushed to the device. Buildozer works only on Linux, and can be downloaded from the given URL https://github.com/kivy/buildozer, then use the command *sudo python setup.py install* to install Buildozer into the system. Finally, it can be used with command *buildozer init.* It will create a buildozer.spec file controlling the build configuration. This is considered as manifested file in Android app. We can set variable to control most or all the parameters passed to python-for-android. *buildozer android debug deploy run* command is used to build, push, and run the application automatically. Some Kivy users may face a problem with it, therefore, some steps to install kivy successfully are given. Steps for Ubuntu are given as follows:

```
$sudo apt install python3
$sudo apt install python3-pip
$sudo apt install python3-dev
$sudo apt install virtualenv
$sudo update-alternatives --config python          #if have more than one python
version
$sudo update-alternatives --install /usr/bin/python3 python /usr/bin/
python3.7 1
$sudo apt install build-essential
$sudo apt install libssl-dev
$sudo apt install libffi-dev
$sudo apt install libxml2-dev
$sudo apt install libxslt1-dev
$sudo apt install zlib1g-dev
$sudo apt install openjdk-8-jdk
$sudo apt install pkg-config
$sudo apt install zlib1g-dev
$sudo apt install libncurses5-dev
$sudo apt install libncursesw5-dev
$sudo apt install libtinfo5
$sudo apt install cmake
$sudo apt install libffi-dev
$sudo nano $HOME/.bashrc
#in the end of file add this line and press ctr+X (^x) then press yes to save
this file modification
export JAVA_HOME=/usr/lib/jvm/java-8-openjdk-amd64
$source $HOME/.bashrc
$sudo apt install git
$sudo apt install libtool
```

```
$sudo apt install cython
$sudo apt install autoconf
$sudo apt install autoreconf
$sudo apt install autotools-dev
$sudo apt install android-sdk
$pip install setuptools
$pip install kivy
$pip install python-for-android
$pip install buildozer
$buildozer init             # to create buildozer.spec, modify as required
$buildozer -v android debug deploy run #takes long time depends upon our computer
```

In case of error Aidl not found, we can install it with the following given command then press "Y" to continue installation:

```
$~/.buildozer/android/platform/android-sdk/tools/bin/sdkmanager  "build-
tools;29.0.0"
```

We can find APK file as an output in the */bin* folder of our Python source code home folder:

```
$buildozer android deploy run logcat  #to launch application on Android device
$buildozer ios deploy run          #to launch application on iOS device
```

We can also set this line at the default command to do if buildozer is stared without any arguments:

```
$buildozer setdefault android debug deploy run logcat > mylog.txt
$buildozer
```

We can share APK file using web server with other non-connected devices. We can browse and download APK file using http://<our_ip_address>:8000

```
$buildozer serve
```

Sample buildozer.spec file is given here, and we can make changes accordingly:

```
[app]

# (str) Title of your application
title = My Application
```

```
# (str) Package name
package.name = myapp

# (str) Package domain (needed for android/ios packaging)
package.domain = org.test

# (str) Source code where the main.py live
source.dir = .

# (list) Source files to include (let empty to include all the files)
source.include_exts = py,png,jpg,kv,atlas

# (list) List of inclusions using pattern matching
#source.include_patterns = assets/*,images/*.png

# (list) Source files to exclude (let empty to not exclude anything)
#source.exclude_exts = spec

# (list) List of directory to exclude (let empty to not exclude anything)
#source.exclude_dirs = tests, bin

# (list) List of exclusions using pattern matching
#source.exclude_patterns = license,images/*/*.jpg

# (str) Application versioning (method 1)
version = 0.1

# (str) Application versioning (method 2)
# version.regex = __version__ = ['"](.*)['"]
# version.filename = %(source.dir)s/main.py

# (list) Application requirements
# comma separated e.g. requirements = sqlite3,kivy
requirements = python3,kivy

# (str) Custom source folders for requirements
# Sets custom source for any requirements with recipes
# requirements.source.kivy = ../../kivy

# (list) Garden requirements
#garden_requirements =

# (str) Presplash of the application
#presplash.filename = %(source.dir)s/data/presplash.png

# (str) Icon of the application
#icon.filename = %(source.dir)s/data/icon.png
```

```
# (str) Supported orientation (one of landscape, sensorLandscape, portrait or
all)
orientation = portrait

# (list) List of service to declare
#services = NAME:ENTRYPOINT_TO_PY,NAME2:ENTRYPOINT2_TO_PY

#
# OSX Specific
#

#
# author = © Copyright Info

# change the major version of python used by the app
osx.python_version = 3

# Kivy version to use
osx.kivy_version = 1.9.1

#
# Android specific
#

# (bool) Indicate if the application should be fullscreen or not
fullscreen = 0

# (string) Presplash background color (for new android toolchain)
# Supported formats are: #RRGGBB #AARRGGBB or one of the following names:
# red, blue, green, black, white, gray, cyan, magenta, yellow, lightgray,
# darkgray, grey, lightgrey, darkgrey, aqua, fuchsia, lime, maroon, navy,
# olive, purple, silver, teal.
#android.presplash_color = #FFFFFF

# (list) Permissions
#android.permissions = INTERNET

# (int) Target Android API, should be as high as possible.
#android.api = 27

# (int) Minimum API your APK will support.
#android.minapi = 21

# (int) Android SDK version to use
#android.sdk = 20
```

```
# (str) Android NDK version to use
#android.ndk = 17c

# (int) Android NDK API to use. This is the minimum API your app will support,
it should usually match android.minapi.
#android.ndk_api = 21

# (bool) Use --private data storage (True) or --dir public storage (False)
#android.private_storage = True

# (str) Android NDK directory (if empty, it will be automatically downloaded.)
#android.ndk_path =

# (str) Android SDK directory (if empty, it will be automatically downloaded.)
#android.sdk_path =

# (str) ANT directory (if empty, it will be automatically downloaded.)
#android.ant_path =

# (bool) If True, then skip trying to update the Android sdk
# This can be useful to avoid excess Internet downloads or save time
# when an update is due, and you just want to test/build your package
# android.skip_update = False

# (bool) If True, then automatically accept SDK license
# agreements. This is intended for automation only. If set to False,
# the default, you will be shown the license when first running
# buildozer.
# android.accept_sdk_license = False

# (str) Android entry point, default is ok for Kivy-based app
#android.entrypoint = org.renpy.android.PythonActivity

# (str) Android app theme, default is ok for Kivy-based app
# android.apptheme = "@android:style/Theme.NoTitleBar"

# (list) Pattern to whitelist for the whole project
#android.whitelist =

# (str) Path to a custom whitelist file
#android.whitelist_src =

# (str) Path to a custom blacklist file
#android.blacklist_src =

# (list) List of Java .jar files to add to the libs so that pyjnius can access
# their classes. Don't add jars that you do not need, since extra jars can slow
# down the build process. Allows wildcards matching, for example:
```

```
# OUYA-ODK/libs/*.jar
#android.add_jars = foo.jar,bar.jar,path/to/more/*.jar

# (list) List of Java files to add to the android project (can be java or a
# directory containing the files)
#android.add_src =

# (list) Android AAR archives to add (currently works only with sdl2_gradle
# bootstrap)
#android.add_aars =

# (list) Gradle dependencies to add (currently works only with sdl2_gradle
# bootstrap)
#android.gradle_dependencies =

# (list) add java compile options
# this can for example be necessary when importing certain java libraries
using the 'android.gradle_dependencies' option
# see https://developer.android.com/studio/write/java8-support for further
information
# android.add_compile_options = "sourceCompatibility = 1.8", "target-
Compatibility = 1.8"

# (list) Gradle repositories to add {can be necessary for some android.
gradle_dependencies}
# please enclose in double quotes
# e.g. android.gradle_repositories = "maven { url 'https://kotlin.bintray.
com/ktor' }"
#android.add_gradle_repositories =

# (list) packaging options to add
# see https://google.github.io/android-gradle-dsl/current/com.android.
build.gradle.internal.dsl.PackagingOptions.html
# can be necessary to solve conflicts in gradle_dependencies
# please enclose in double quotes
# e.g. android.add_packaging_options = "exclude 'META-INF/common.kotlin_-
module'", "exclude 'META-INF/*.kotlin_module'"
#android.add_gradle_repositories =

# (list) Java classes to add as activities to the manifest.
#android.add_activites = com.example.ExampleActivity

# (str) OUYA Console category. Should be one of GAME or APP
# If you leave this blank, OUYA support will not be enabled
#android.ouya.category = GAME
```

```
# (str) Filename of OUYA Console icon. It must be a 732x412 png image.
#android.ouya.icon.filename = %(source.dir)s/data/ouya_icon.png

# (str) XML file to include as an intent filters in <activity> tag
#android.manifest.intent_filters =

# (str) launchMode to set for the main activity
#android.manifest.launch_mode = standard

# (list) Android additional libraries to copy into libs/armeabi
#android.add_libs_armeabi = libs/android/*.so
#android.add_libs_armeabi_v7a = libs/android-v7/*.so
#android.add_libs_arm64_v8a = libs/android-v8/*.so
#android.add_libs_x86 = libs/android-x86/*.so
#android.add_libs_mips = libs/android-mips/*.so

# (bool) Indicate whether the screen should stay on
# Don't forget to add the WAKE_LOCK permission if you set this to True
#android.wakelock = False

# (list) Android application meta-data to set (key=value format)
#android.meta_data =

# (list) Android library project to add (will be added in the
# project.properties automatically.)
#android.library_references =

# (list) Android shared libraries which will be added to AndroidManifest.xml
using <uses-library> tag
#android.uses_library =

# (str) Android logcat filters to use
#android.logcat_filters = *:S python:D

# (bool) Copy library instead of making a libpymodules.so
#android.copy_libs = 1

# (str) The Android arch to build for, choices: armeabi-v7a, arm64-v8a, x86,
x86_64
android.arch = armeabi-v7a

#
# Python for android (p4a) specific
#

# (str) python-for-android fork to use, defaults to upstream (kivy)
#p4a.fork = kivy
```

```
# (str) python-for-android branch to use, defaults to master
#p4a.branch = master

# (str) python-for-android git clone directory (if empty, it will be automati-
cally cloned from github)
#p4a.source_dir =

# (str) The directory in which python-for-android should look for your own
build recipes (if any)
#p4a.local_recipes =

# (str) Filename to the hook for p4a
#p4a.hook =

# (str) Bootstrap to use for android builds
# p4a.bootstrap = sdl2

# (int) port number to specify an explicit --port= p4a argument (eg for boot-
strap flask)
#p4a.port =

#
# iOS specific
#

# (str) Path to a custom kivy-ios folder
#ios.kivy_ios_dir = ../kivy-ios
# Alternately, specify the URL and branch of a git checkout:
ios.kivy_ios_url = https://github.com/kivy/kivy-ios
ios.kivy_ios_branch = master

# Another platform dependency: ios-deploy
# Uncomment to use a custom checkout
#ios.ios_deploy_dir = ../ios_deploy
# Or specify URL and branch
ios.ios_deploy_url = https://github.com/phonegap/ios-deploy
ios.ios_deploy_branch = 1.7.0

# (str) Name of the certificate to use for signing the debug version
# Get a list of available identities: buildozer ios list_identities
#ios.codesign.debug = "iPhone Developer: <lastname><firstname> (<hexstring>)"

# (str) Name of the certificate to use for signing the release version
#ios.codesign.release = %(ios.codesign.debug)s

[buildozer]
```

```
# (int) Log level (0 = error only, 1 = info, 2 = debug (with command output))
log_level = 2

# (int) Display warning if buildozer is run as root (0 = False, 1 = True)
warn_on_root = 1

# (str) Path to build artifact storage, absolute or relative to spec file
# build_dir = ./.buildozer

# (str) Path to build output (i.e. .apk, .ipa) storage
# bin_dir = ./bin

#--------------------------------------------------------------------------
# List as sections
#
# You can define all the "list" as [section:key].
# Each line will be considered as a option to the list.
# Let's take [app] / source.exclude_patterns.
# Instead of doing:
#
#[app]
#source.exclude_patterns = license,data/audio/*.wav,data/images/original/
*
#
# This can be translated into:
#
#[app:source.exclude_patterns]
#license
#data/audio/*.wav
#data/images/original/*
#


#--------------------------------------------------------------------------
# Profiles
#
# You can extend section / key with a profile
# For example, you want to deploy a demo version of your application without
# HD content. You could first change the title to add "(demo)" in the name
# and extend the excluded directories to remove the HD content.
#
#[app@demo]
#title = My Application (demo)
```

```
#
#[app:source.exclude_patterns@demo]
#images/hd/*
#
# Then, invoke the command line with the "demo" profile:
#
#buildozer --profile demo android debug
```

A Dockerfile is available to use buildozer through a Docker environment with the following command:

```
$docker build --tag=buildozer .
$docker run --volume "$(pwd)":/home/user/hostcwd buildozer --version
```

Sample Docker file is given as:

```
# Dockerfile for providing buildozer
#
# Build with:
# docker build --tag=kivy/buildozer.
#
# In order to give the container access to your current working directory
# it must be mounted using the --volume option.
# Run with (e.g. `buildozer --version`):
# docker run \
# --volume "$HOME/.buildozer":/home/user/.buildozer
# --volume "$PWD":/home/user/hostcwd
# kivy/buildozer --version
#
# Or for interactive shell:
# docker run --interactive --tty --rm \
# --volume "$HOME/.buildozer":/home/user/.buildozer
# --volume "$PWD":/home/user/hostcwd
# --entrypoint /bin/bash
# kivy/buildozer
#
# If you get a `PermissionError` on `/home/user/.buildozer/cache`,
# try updating the permissions from the host with:
# sudo chown $USER -R ~/.buildozer
# Or simply recreate the directory from the host with:
# rm -rf ~/.buildozer && mkdir ~/.buildozer
```

```
FROM ubuntu:18.04

ENV USER="user"
ENV HOME_DIR="/home/${USER}"
ENV WORK_DIR="${HOME_DIR}/hostcwd" \
  SRC_DIR="${HOME_DIR}/src" \
  PATH="${HOME_DIR}/.local/bin:${PATH}"

# configures locale
RUN apt update -qq > /dev/null &&
  apt install -qq --yes --no-install-recommends \
  locales && \
  locale-gen en_US.UTF-8
ENV LANG="en_US.UTF-8" \
  LANGUAGE="en_US.UTF-8" \
  LC_ALL="en_US.UTF-8"

# system requirements to build most of the recipes
RUN apt install -qq --yes --no-install-recommends \
  autoconf \
  automake \
  build-essential \
  ccache \
  cmake \
  gettext \
  git \
  libffi-dev \
  libltdl-dev \
  libtool \
  openjdk-8-jdk \
  patch \
  pkg-config \
  python2.7 \
  python3-pip \
  python3-setuptools \
  sudo \
  unzip \
  zip \
  zlib1g-dev

# prepares non root env
RUN useradd --create-home --shell /bin/bash ${USER}
# with sudo access and no password
RUN usermod -append --groups sudo ${USER}
```

```
RUN echo "%sudo ALL=(ALL) NOPASSWD: ALL" >> /etc/sudoers

USER ${USER}
WORKDIR ${WORK_DIR}
COPY --chown=user:user . ${SRC_DIR}

# installs buildozer and dependencies
RUN pip3 install --user Cython==0.28.6 ${SRC_DIR}

ENTRYPOINT ["buildozer"]
```

12.1.4 Plyer

Plyer is a platform-independent API to use features commonly found on various platforms. It provides APIs for devices for sensors such as accelerometer, camera, compass, GPS, screen brightness, and notifications. These APIs are useful in all devices such as desktop and mobile. We will work on Plyer version 1.4.2. It can be installed with the help of the following command:

```
$pip install plyer
```

After doing all this we can build the app with the help of buildozer, and as a result, we can perform all the hardware-related operations. The player also supported by Raspberry Pi. We are not including it for compatibility, but we can expect it shortly. Compatible sheet for Plyer is given in Table 12.1:

Table 12.1: Various hardware compatibilities.

Platform	Windows	OS X	Linux	Android	iOS
Accelerometer	No	Yes	Yes	Yes	Yes
Audio recording	Yes	Yes	No	Yes	No
Barometer	No	No	No	Yes	Yes
Battery	Yes	Yes	Yes	Yes	Yes
Bluetooth	No	Yes	No	Yes	No
Brightness	No	No	Yes	Yes	Yes
Call	No	No	No	Yes	Yes
Camera (taking picture)	No	No	No	Yes	Yes
Compass	No	No	No	Yes	Yes
CPU count	Yes	Yes	Yes	No	No

Table 12.1: (continued)

Platform	Windows	OS X	Linux	Android	iOS
Email (open mail client)	Yes	Yes	Yes	Yes	Yes
Flash	No	No	No	Yes	Yes
GPS	No	No	No	Yes	Yes
Gravity	No	No	No	Yes	Yes
Gyroscope	No	No	No	Yes	Yes
Humidity	No	No	No	Yes	No
IR Blaster	No	No	No	Yes	No
Light	No	No	No	Yes	No
Native file chooser	Yes	Yes	Yes	Yes	No
Notifications	Yes	Yes	Yes	Yes	No
Orientation	No	No	No	Yes	No
Proximity	No	No	No	Yes	No
Screenshot	Yes	Yes	Yes	No	No
SMS (send messages)	No	No	No	Yes	Yes
Spatial orientation	No	No	No	Yes	Yes
Speech to text	No	No	No	Yes	No
Storage Path	Yes	Yes	Yes	Yes	Yes
Temperature	No	No	No	Yes	No
Text to speech	Yes	Yes	Yes	Yes	Yes
Unique ID	Yes	Yes	Yes	Yes	Yes
Vibrator	No	No	No	Yes	Yes
Wifi	Yes	Yes	Yes	No	No

12.2 Building an android APK on OS X

We can build an app on OS X with Python3. A Kivy application can be exported for any type of platform. Android is one of the major and popular platforms. It will provide a normal and fully standalone APK file just like we create from Java application. This APK file can be published over the Google Play store too. Many tools are available to do so. For more information, we can refer to its official documentation.

We can package directly with python-for-android, which gives more control, but it requires us to manually download parts of the Android toolchain. We need to initialize the configuration file (*buildozer.spec*) by using the command *buildozer init*. Let us see how to build app on OS X:

```
$ pip install buildozer                          #to install buildozer using pip
$ sudo pip install https://github.com/kivy/buildozer/archive/master.zip
                                   #latest version from repository
$ brew install autoconf automake
$ pip install colorama appdirs sh jinja2 six
```

12.3 Package licensing

We can write our licensed APK file for releasing purpose in this channel buildozer requires Keystore information. We can set these values as environment variables. We need to provide Keystore path, Keystore password, key alias, and key password:

```
export P4A_RELEASE_KEYSTORE=$HOME/.keystore
export P4A_RELEASE_KEYSTORE_PASSWD=aaaAAA123
export P4A_RELEASE_KEYALIAS_PASSWD=aaaAAA123
export P4A_RELEASE_KEYALIAS=mykeystore
```

The generated file will be assigned APK, which can directly be published over the Google Play Store. Now, we can build APK file with this command:

```
$python3 -m buildozer -v android release
or
$buildozer -v android release
```

12.4 Packaging application using Kivy launcher

Kivy launcher is an Android application that runs any Kivy example stored on our SD card. To install the Kivy launcher, we must visit Kivy launcher page on Google Play Store then install it by selecting our android phone or APK can be installed in our phone directly from this URL http://kivy.org/#download

12.5 External libraries

Kivy comes with some predefined external libraries from Python or C Language. Such as *ddsfile*, which is used to parse and save DDS files, *mydev* is used for kernel multi-touch transformation library and *osc* is used to open sound control protocols. The osc is an upgraded version of *PyOSC* library. GstPlayer is an external library and Kivy does not provide any support for it. It might change in the future and we should not rely on it for our code. GstPlayer is a media player implemented specifically for Kivy with gstreamer 1.0. GstPlayer is automatically compiled if we have pkg-config-libs-cflags gstreamer-1.0 working. Kivy is not responsible for the libraries. It does not use Gi at all and is focused on what we want: the ability to read video and stream the image in a callback or read an audio file. We do not use it directly but use our core provider instead:
GstPlayer.

12.6 Python application testing

The Kivy and Python being open source, they provide lots of tools for testing purpose. Some of the tools are from the Python application test (pytest, unittest) and some are specialized for Kivy application (telenium) only. All the given codes in this book are valid for both versions of Python, but for the best results, tested environment is Python 3.7. As we know, App class in Kivy is considered as the main entry point into Kivy run loop. We have to create an instance of our App child class then call run function. In Kivy application testing, a widely used library is unittest. Unittest tests are expressive, more readable, easy and without any boiler plate code. It is referred to as "PyUnit" or Python version of JUnit. Unittest will prevent bugs from coming back unnoticed in the future. GL context with image compression (GUI) and command line-based testing (CLI) are the two types of unit testing. Study an example by PyUnit:

```python
import unittest

class TestStringMethods(unittest.TestCase):

    def test_upper(self):
        self.assertEqual('tarkeshwar'.upper(), 'TARKESHWAR')

    def test_isupper(self):
        self.assertTrue("TARKESHWAR".isupper())
        self.assertFalse("Tarkeshwar".isupper())

    def test_split(self):
        s = 'Tarkeshwar Barua'
        self.assertEqual(s.split(), ['Tarkeshwar', 'Barua'])
        with self.assertRaises(TypeError):
```

```
        s.split(2)

if __name__ == '__main__':
  unittest.main()
--------------------------------output--------------------------------
$python pyunitTestdemo.py
. . .
----------------------------------------------------------------------
Ran 3 tests in 0.001s
OK
----------------------------------------------------------------------
```

Only three Python results are possible where OK means tests are passed without any error, FAIL means test does not pass and raise an AssertionError exception (symbol A), and ERROR means test raises an exception other than AssertionError (symbol E). In any unit testing, there are some terms being used. The terms are as follows:

Test fixture – It represents the preparation needed to perform one or more tests and any associate cleanup actions.

Test case – It is the smallest unit of testing where it checks for a specific response to a set of code. TestCase is a base class in unittest, which helps to create new test cases.

Test suite – A collection of test cases and test suites is known as a test suite. It is the process to combine more than one test as a single unit for a test.

Test runner – It is used for running test cases and provides the outcome to the user. It may be graphical and command line which returns a value.

Pytest is a framework that makes building a simple and scalable test easy. Here, Python pytest version 5.3.2 is used. Create one folder named with a test at home directory of the app and install pytest library using pip as given in the following code:

```
$pip install pytest
$pytest --version
This is pytest version 5.3.2, imported from c:\users\tarkeshwar\appdata
\local\programs\python\python37\lib\sit
e-packages\pytest\__init__.py

$pytest
============================ test session starts =====================
platform win32 -- Python 3.7.6rc1, pytest-5.3.2, py-1.8.1, pluggy-0.13.1
rootdir: C:\Users\Tarkeshwar\PycharmProjects\KiVyPrograms
collected 0 items
============================ no tests ran in 0.60s ==================
```

Let us see some sample code (sample.py) to test:

```
def squareTest(x):
    return x * x

def test_answer():
    assert squareTest(3) == 9
```

The above given code can be tested with the following command:

```
$pytest one.py
============================== test session starts =========================
platform win32 -- Python 3.7.6rc1, pytest-5.3.2, py-1.8.1, pluggy-0.13.1
rootdir: C:\Users\Tarkeshwar\PycharmProjects\KiVyPrograms\kivyApplication
Testing
collected 1 item

one.py . [100%]

============================== 1 passed in 0.09s ==========================

$pytest one.py
============================== test session starts =========================
platform win32 -- Python 3.7.6rc1, pytest-5.3.2, py-1.8.1, pluggy-0.13.1
rootdir: C:\Users\Tarkeshwar\PycharmProjects\KiVyPrograms\kivyApplication
Testing
collected 1 item

one.py F = [100%]

============================== FAILURES ================================
_____ test_answer _____
    def test_answer():
>       assert inc(3) == 5
E       assert 4 == 5
E       + where 4 = inc(3)

one.py:6: AssertionError
============================== 1 failed in 0.14s =========================
```

When kivy app starts, it creates a loop and until the loop continues nothing will be executed after *App().run()*, that is why we have to break the loop. We can break the loop with the help of *time.sleep()* that will pause our application for a while. Let us see the sample code for test:

```python
from kivy.app import App
from kivy.lang import Builder
from kivy.uix.boxlayout import BoxLayout
from kivy.uix.button import Button

Builder.load_string('''
<MyButton>:
    text: 'Click Me'
    on_release: print('Button Has been clicked!')
<Body>:
    MyButton:
''')
class MyButton(Button):
    def __init__(self, **kwargs):
            super(MyButton, self).__init__(**kwargs)
            app = App.get_running_app()
            app.my_button = self

class Body(BoxLayout):
    pass

class My(App):
    def build(self):
        return Body()

if __name__ == '__main__':
    My().run()
```

Figure 12.9: Sample Kivy app for testing.

Given is an example of a kivy application test code:

```python
import unittest
import time
from functools import partial
from kivy.clock import Clock
from apptesting.Main import My

class Test(unittest.TestCase):
    def pause(*args):
        time.sleep(0.000001)

    def run_test(self, app, *args):
        Clock.schedule_interval(self.pause, 0.000001)
        app.my_button.dispatch('on_release')
        self.assertEqual('Click Me', app.my_button.text)
        app.stop()

    def test_example(self):
        app = My()
        p = partial(self.run_test, app)
        Clock.schedule_once(p, 0.000001)
        app.run()

if __name__ == '__main__':
    unittest.main()
```

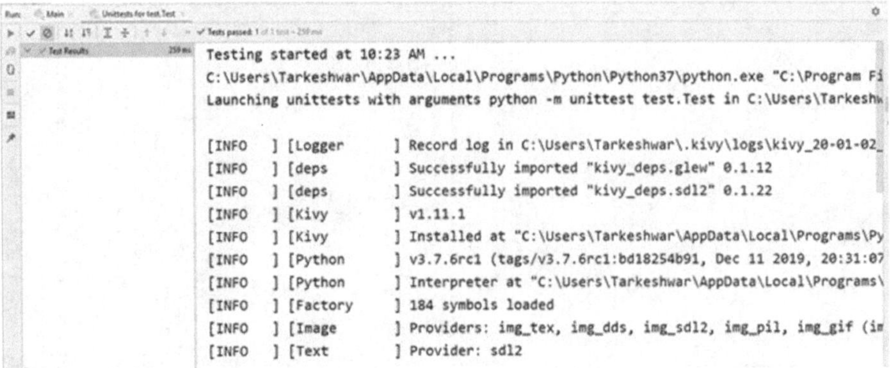

Figure 12.10: Kivy app testing result.

12.7 Release on the market

Kivy is supported by multitouch devices. It can run on Android phones using the Kivy Launch. Kivy applications can be published on the Android play store. After building, we can create a release version. We must run buildozer with the release parameter, or if using python-for-android we use the release option to build.py. This creates releasing APK in the bin folder, which must be properly signed by us and for this, follow the steps given on the URL https://developer.android.com/stu dio/publish/app-signing.html#signing-manually

12.8 Kivy application testing

The nose was written by Jason Pellerin in 2005 to support the same test by *py.test*. Kivy comes with nosetest, and nose is one of the major testing libraries that can be installed using the following command:

```
$ pip install nose      #to install nose
$ make test             #to run test suit

#module name noseTest.py
def square(x):
    return x * x

def test_answer():
    assert square(10) == 100

$nosetests noseTest.py
----------------------------------------------------------------------
Ran 1 test in 0.001s

OK
```

Summary

- The behavior can be added locally and globally with the help of BehaviorMixin. The idea behind these classes is to encapsulate properties and events associated with the type of widgets. This module implements behaviors that can be mixed in with existing widgets.
- Packaging could be either 32 bit or 64 bit, which depends on the Python version and platforms such as Windows, Linux, Android, and iOS. PyInstaller is one of the tools that is used for packing.

- Pyinstaller 3.5. uses *spec* file for configuration for the packaging app. The executable file generation process takes some time. After completion of this process, we can find executable file in dist folder.
- Python4 Android should be at the class path to the Python for Android installation, and the current directory is called "testapp," the given script after the application code.
- The testing device could be any physical and virtual devices.
- The default orientation is landscape, but it can be changed by using the command "--orientation portrait," and icon can be replaced with "--icon $currdir/test_icon.png."
- Before releasing APK over Play Store, we need to sign it first.
- Python-for-android tool is a popular packaging tool for Android. It is only supported by Linux. Python-for-android supports both Python2 and Python3 versions. We can package our fully optimized app for multiple architectures.
- Buildozer is one of the tools that automates this entire process. The easiest way to Android package is Buildozer to make full APK. We can also run our app without a compilation step with Kivy Launcher. These apps can be released over all the available platforms as signed or unsigned app.
- Dockerfile can be used to build kivy app with buildozer.
- The Plyer provides APIs for devices for sensors such as accelero meter, camera, compass, GPS, screen brightness, and notifications. These APIs are useful in all devices such as desktop and mobile.
- We can package directly with Python-for-android, which can give more control but it requires us to manually download parts of the Android tool chain. We need to initialize the configuration file (buildozer.spec) by using the command buildozer init.
- We can write our licensed APK file for releasing purpose in this channel buildozer requires keystore information. We can set these values as environment variables. We need to provide keystore path, keystore password, key alias, and key password.
- Kivy launcher is an Android application that runs any Kivy example stored on our SD card.
- GstPlayer is a media player implemented specifically for Kivy with gstreamer 1.0. GstPlayer is automatically compiled if we have pkg-config-libs-cflags gstreamer-1.0 working. Kivy is not responsible for the libraries.
- We can test our Kivy application using pytest, unittest, nose, and some tests that are specialized for Kivy application (telenium) only.
- The possible application test results are OK, A, and E, where OK means tests are passed without any error, FAIL means test does not pass and raise an AssertionError exception (symbol A), and ERROR means test raises an exception other than AssertionError (symbol E).

– We can break loop with the help of time.sleep() that will pause our application for a while and only then the application can be tested.
– We can build a release version, we must run buildozer with the release parameter, or if using Python-for-android we use the – release option to build.py. This creates release APK in the bin folder.
– The nose was written by Jason Pellerin in 2005 to support the same test by py. test. Kivy comes with nosetest, and nose is one of the major testing libraries.

Key terms

The ButtonBehavior mixin class can be combined with widgets, such as an Image, to provide alternative buttons that preserve Kivy button behavior. PyInstaller is one of the tools that is used for packing Kivy application for windows platform and buildozer and Python-for Android for Android, Linux, iOS, and Mac platforms. This whole build process can be controlled by *buildozer.spec* file and the result can be found in dist and bin folder. While we are testing an app with Android then Developer Option must be enabled. The process may take some time to generate .APK, and installation of our app depends on the computer performance. Before releasing APK over Play Store, APK must be signed. Python-for-android tool is a popular packaging tool for Android. It is only supported by Linux. Python-for-android supports both Python2 and Python3 versions. We can package our fully optimized app for multiple architectures but Buildozer is one of the tools that automates this entire process. The easiest way to Android package is Buildozer to make full APK. We can also run our app without a compilation step with Kivy Launcher. We can share APK files using a web server to other non-connected devices. APK file can be shared, browsed, and downloaded using buildozer command *buildozer serve* command, and it can be accessedt with the help of http://<our_ip_address>:8000/. The buildozer can be configured over docker using docker file. The Plyer provides APIs for devices for sensors such as accelero meter, camera, compass, GPS, screen brightness, and notifications. These APIs are useful in all devices such as desktop and mobile. The player is also supported by Raspberry Pi. We can build app on OS X with Python3. A Kivy application can be exported for any type of platform. Android is one of the major and popular platforms. It will provide a normal and fully standalone APK file just like we create from Java application. We can build license APK by providing keystore value information, which can be set as environment variables. We need to provide keystore path, keystore password, key alias, and key password. The resulting file can be uploaded on the Google Play Store. Kivy launcher is an Android application that runs any Kivy example stored on our SD card. Kivy comes with some predefined external libraries from Python or C Language. Such as ddsfile, which is used to parse and save DDS files, mydev is used for kernel multi-touch transformation library, and osc is used to open sound control protocols. The *unittest* tests are expressive, more readable, easy and without any boiler plate code. It is referred to as "PyUnit" or Python version of JUnit. Unittest will prevent bugs from coming back unnoticed in the future. GL context with image compression (GUI) and command line-based testing (CLI) are the two types of unit testing. The terminologies used in application testing are test fixture, test cases, test suite, and test runner. The Pytest is another framework that makes building simple and scalable test easy. When kivy app starts, it creates a loop and until the loop continues, nothing will execute after App().run() that is why we have to break the loop. We can break a loop with the help of time.sleep() that will pause our application for a while. The nose supports the same test by py.test. Kivy comes with nosetest, nose is one of the major testing libraries.

Review questions

1. Which module is responsible for adding behavior to the UI component?
2. List all the packaging tools for various platforms.
3. What does PyInstaller do?
4. What is the role of .spec file in packaging any Python and Kivy app?
5. How do we enable Developer Options in Android devices?
6. What is the role of Python-for-android in building an app?
7. How do we generate signed APK to upload on Google Play Store?
8. What are the components to generate signed APK using Python-for-android?
9. What is buildozer?
10. What is the role of buildozer.spec in packaging app?
11. What will be the output folder of buildozer build package?
12. What is Aidl and how to resolve it (Aidl not found)?
13. How do we share APK file over network?
14. How do we deploy buildozer using Dockerfile?
15. What is the use of Plyer?

Exercise

Tick the correct option

Q.1. Which type of APK file can be published over Google Play Store _____?
    a) Unassigned
    b) Unsigned
    c) Signed
    d) Assigned

Q.2. Which tool is not used to package application in OS X _____?
a) Python-for-android
b) Buildozer
c) Kivy launcher
d) Pytest

Q.3. Which of these is buildozer configuration file _____?
    a) buildozer.spec   b) buildozer.conf   c) buildozer.ini   d) buildozer.txt

Q.4. What is the key component of key store _____?
a) Key password
b) keystore password
c) key store path
d) All of the above

Q.5. What is Kivy launcher _____?
a) Android application
b) Python library
c) API
d) .jar file

Q.6. Kivy can use which type of library _____?
a) jar files
b) c libraries
c) class files
d) All of the above

Q.7. Which libraries cannot be used in Kivy application _____?
a) Junit
b) Pytest
c) Unittest
d) Telenium

Q.8. Which one is not a possible result in application testing _____?
a) A
b) OK
c) Okey
d) E

Q.9. What is test suite _____?
a) It represents the preparation needed to perform one or more tests, and any associate cleanup actions.

b) A collection of test cases and test suites is known as a test suite. It is the process to combine more than one test as a single unit for test.

c) It is used for running test cases and provides the outcome to the user. It may be graphical and command line, which returns a value.

d) It is the smallest unit of testing where it checks for a specific response to a particular set of code. TestCase is base class in unittest which helps to create new test cases.

Q.10. Who wrote the nose _____?
    a) Charles Babbage
    b) Dennis Richie
    c) James Goslin
    d) Jason Pellerin

Answers

Q.1. c) Signed
Q.2. d) Pytest
Q.3. a) buildozer.spec
Q.4. d) all of the above
Q.5. a) Android application
Q.6. d) All of the above
Q.7. a) JUnit
Q.8. c) Okey
Q.9. b) A collection of test cases and test suites is known as test suite. It is the process to combine more than one test as a single unit for test.
Q.10. d) Jason Pellerin

Fill in the blanks

1. The idea behind these classes is to _____ and _____ associated with type of widgets.
2. _____ is one of the tools used for packing in windows.
3. The executable file can be found in _____ folder.
4. _____ is a tool that automates the entire process.
5. APK file as an output in the _____ folder of our Python source code home folder.

Answers

1. encapsulate properties, events
2. PyInstaller
3. /dist
4. buildozer
5. /bin

13 Sample project

13.1 Introduction of git

Software development is not an easy task, and it becomes very difficult while we are handling a big code and multiple programmers are working on the same project. Decision-making is needed, whether updated code should be a part of the production environment. Any change can be accepted or rejected at any phase of the development and releasing process in the environment. Managing code in this is called version controlling, and to do so, applications such as git, SVN, TFS, and so on, are used. Git is a very popular and open-source application utility. Git is developed by Linux Kernel by developer Linus Torvalds. There are plenty of git implementations available such as http://www.gitgub.com, https://about.gitlab.com/, https://bitbucket.org/product/, and many more. Most of the git distributions are free of cost and available with public repository, but some of them charge a small amount to provide a private repository.

13.2 Implementation of git

Version Control System (VCS) is a software that helps software developers to work together and maintain a complete history of their work. Some of the functions are listed as follows:

- It allows developers to work simultaneously.
- It does not allow overwriting each other's changes.
- Maintains a history of every version.

We can download git from https://git-scm.com/downloads for various platforms. As we can see in Figure 13.1.

It provides a command line tool as well as a GUI tool. In Linux (Ubuntu) we can install using *sudo apt install git* command. Some of the git commands are given as follows:

```
$ git config --global user.name "Name Surname"
$ git config --global user.email name.surname@gmail.com
$ git clone username@192.168.10.10:/usr/local/cryo3d/cryo3d.git
$ git add filename [start tracking new/edited filename]
$ git add . (git add -A)[start tracking all changed/new files]
$ git commit -m 'Commit message: what changes were introduced'
$ git add sort.c# adds file to the staging area[bash]
```

https://doi.org/10.1515/9783110689488-013

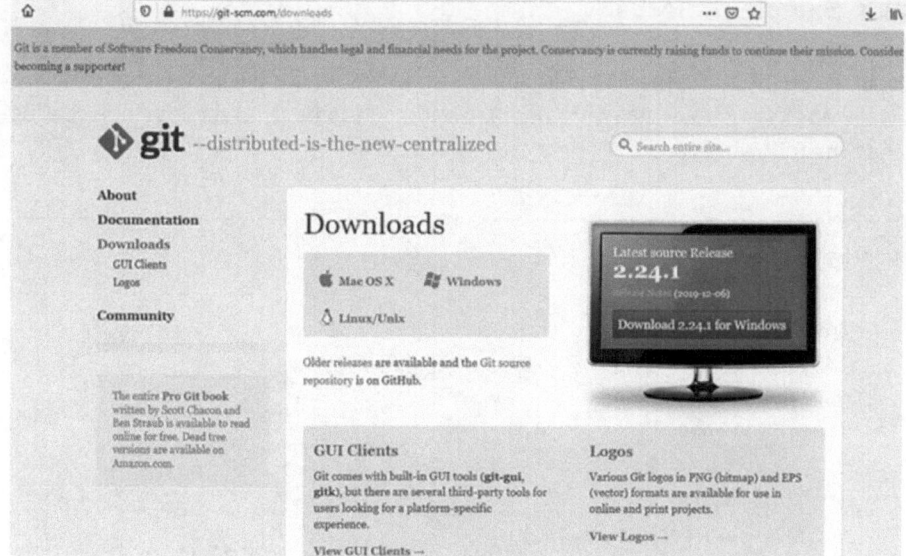

Figure 13.1: Git home page.

Figure 13.2: Git download page.
Source: https://git-scm.com/download/win

```bash
$ git commit —m "Addedsort operation"# Second commit
$ git add search.c# adds file to the staging area$ git commit —m
"Addedsearch operation"
$ git status
```

13.3 Sample project 1

In the given program, the GUI components and their working with internal code are displayed. Kv language is a very popular approach to develop the kivy app. Let us study the following code:

```
-----------------------------------sample.kv----------------------------
<CustLabel@Label>:
    color: 0, 0, 0, 1
<CustomPopup>:
    size_hint: .5, .5
    auto_dismiss: False
    title: "The Popup"
    Button:
        text: "Close"
        on_press: root.dismiss()
SampBoxLayout:
<SampBoxLayout>:
    orientation: "vertical"
    padding: 10
    spacing: 10
    # ---------- Holds CheckBox and RadioBox ----------
    BoxLayout:
        orientation: "horizontal"
        height: 30
        BoxLayout:
            orientation: "horizontal"
            size_hint_x: .22
            CustLabel:
                text: "Are you over 18"
                size_hint_x: .80
            CheckBox:
                on_active: root.checkbox_18_clicked(self, self.active)
                size_hint_x: .20
        BoxLayout:
            orientation: "horizontal"
            size_hint_x: .55
            CustLabel:
                text: "Favorite Color"
                color: 0, 0, 0, 1
                size_hint_x: .265
```

```
            CheckBox:
                group: "fav_color"
                value: root.blue
                size_hint_x: .05
            CustLabel:
                text: "Blue"
                color: 0, 0, 0, 1
                size_hint_x: .15
            CheckBox:
                group: "fav_color"
                value: root.red
                size_hint_x: .05
            CustLabel:
                text: "Red"
                color: 0, 0, 0, 1
                size_hint_x: .15
            CheckBox:
                group: "fav_color"
                value: root.green
                size_hint_x: .05
            CustLabel:
                text: "Green"
                color: 0, 0, 0, 1
                size_hint_x: .15
# ---------- Holds Slider & Switch ----------
BoxLayout:
    orientation: "horizontal"
    height: 30
    BoxLayout:
        orientation: "horizontal"
        size_hint_x: .25
        CustLabel:
            text: str(slider_id.value)
        # Define the min, max, starting value and how
        # much the value changes with each move
        Slider:
            id: slider_id
            min: -100
            max: 100
            value: 0
            step: 1
```

```
    BoxLayout:
        orientation: "horizontal"
        size_hint_x: .25
        CustLabel:
            text: "On / Off"
        Switch:
            id: switch_id
            on_active: root.switch_on(self, self.active)
# ---------- Displays Popup & Spinner ----------
BoxLayout:
    orientation: "horizontal"
    height: 30
    BoxLayout:
        orientation: "horizontal"
        size_hint_x: .25
        # When clicked the popup opens
        Button:
            text: "Open Popup"
            on_press: root.open_popup()
    BoxLayout:
        orientation: "horizontal"
        size_hint_x: .25
        Spinner:
            text: "First"
            values: ["First", "Second", "Third"]
            id: spinner_id
            on_text: root.spinner_clicked(spinner_id.text)
# ---------- Displays TabbedPanel ----------
BoxLayout:
    orientation: "horizontal"
    height: 30
    BoxLayout:
        orientation: "horizontal"
        size_hint_x: .25
        TabbedPanel:
            do_default_tab: False
            TabbedPanelItem:
                text: "1st Tab"
                Label:
                    text: "Content of First Panel"
            TabbedPanelItem:
                text: "2nd Tab"
```

```
            Label:
                    text: "Content of Second Panel"
            TabbedPanelItem:
                    text: "3rd Tab"
                    Label:
        text: "Content of Third Panel"
```

```python
import kivy
kivy.require("1.9.0")
from kivy.app import App
from kivy.uix.boxlayout import BoxLayout
from kivy.properties import ObjectProperty
from kivy.core.window import Window
from kivy.uix.popup import Popup
# Used to display popup
class CustomPopup(Popup):
    pass
class SampBoxLayout(BoxLayout):
    # For checkbox
    checkbox_is_active = ObjectProperty(False)
    def checkbox_18_clicked(self, instance, value):
        if value is True:
            print("Checkbox Checked")
        else:
            print("Checkbox Unchecked")
    # For radio buttons
    blue = ObjectProperty(True)
    red = ObjectProperty(False)
    green = ObjectProperty(False)
    # For Switch
    def switch_on(self, instance, value):
        if value is True:
            print("Switch On")
        else:
            print("Switch Off")
    # Opens Popup when called
    def open_popup(self):
        the_popup = CustomPopup()
        the_popup.open()
    # For Spinner
    def spinner_clicked(self, value):
```

```
            print("Spinner Value " + value)
class SampleApp(App):
    def build(self):
            # Set the background color for the window
            Window.clearcolor = (1, 1, 1, 1)
            return SampBoxLayout()
sample_app = SampleApp()
sample_app.run()
```

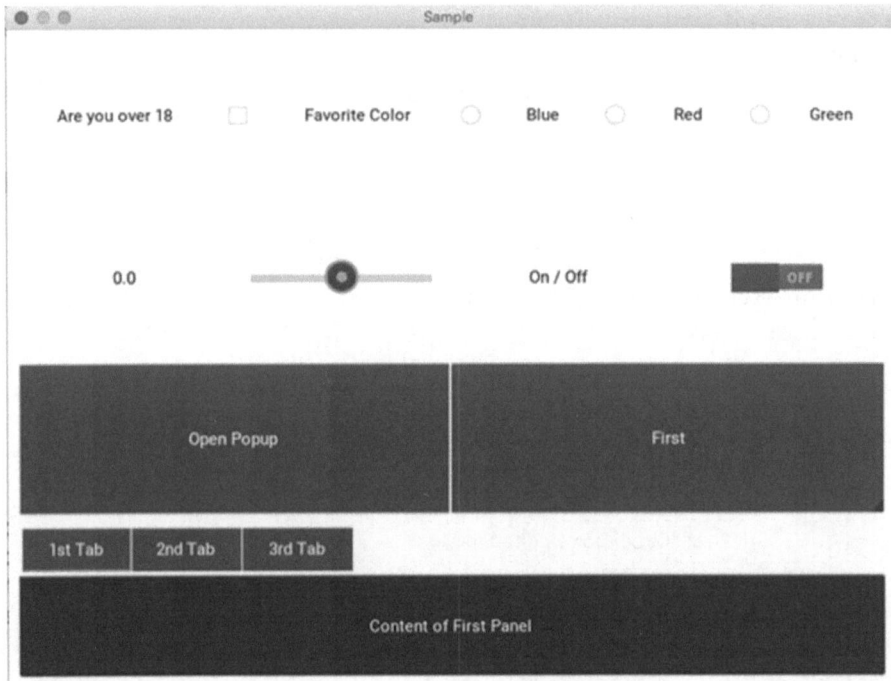

Figure 13.3: First project.

Sample project 2

The camera is a very important part of the device, and we can do many tasks with it. In the given project, we are going to shuffle the camera view in various blocks:

```
from kivy.app import App
from kivy.uix.camera import Camera
from kivy.uix.widget import Widget
from kivy.uix.slider import Slider
```

```python
from kivy.uix.scatter import Scatter
from kivy.animation import Animation
from kivy.graphics import Color, Rectangle
from kivy.properties import NumericProperty
from random import randint, random
from functools import partial

class Puzzle(Camera):
    blocksize = NumericProperty(100)
    def on_texture_size(self, instance, value):
        self.build()

    def on_blocksize(self, instance, value):
        self.build()

    def build(self):
        self.clear_widgets()
        texture = self.texture
        if not texture:
            return
        bs = self.blocksize
        tw, th = self.texture_size
        for x in range(int(tw / bs)):
            for y in range(int(th / bs)):
                bx = x * bs
                by = y * bs
                subtexture = texture.get_region(bx, by, bs, bs)
                # node = PuzzleNode(texture=subtexture,
                # size=(bs, bs), pos=(bx, by))
                node = Scatter(pos=(bx, by), size=(bs, bs))
                with node.canvas:
                    Color(1, 1, 1)
                    Rectangle(size=node.size, texture=subtexture)
                self.add_widget(node)
        self.shuffle()

    def shuffle(self):
        texture = self.texture
        bs = self.blocksize
        tw, th = self.texture_size
        count = int(tw / bs) * int(th / bs)
        indices = list(range(count))
        childindex = 0
        while indices:
```

```
                    index = indices.pop(randint(0, len(indices) - 1))
                    x = bs * (index % int(tw / bs))
                    y = bs * int(index / int(tw / bs))
                    child = self.children[childindex]
                    a = Animation(d=random() / 4.) + Animation(pos=(x, y), t='out_-
                    quad', d=.4)
                    a.start(child)
                    childindex += 1

        def on_touch_down(self, touch):
            if touch.is_double_tap:
                self.shuffle()
                return True
            super(Puzzle, self).on_touch_down(touch)

class PuzzleApp(App):
    def build(self):
        root = Widget()
        puzzle = Puzzle(resolution=(640, 480), play=True)
        slider = Slider(min=100, max=200, step=10, size=(800, 50))
        slider.bind(value=partial(self.on_value, puzzle))
        root.add_widget(puzzle)
        root.add_widget(slider)
        return root

    def on_value(self, puzzle, instance, value):
        value = int((value + 5) / 10) * 10
        puzzle.blocksize = value
        instance.value = value

PuzzleApp().run()
```

Figure 13.4: Camera suffeler.

13.4 Kivy catalog

Kivy catalog viewer showcase widgets are available in Kivy and allow interactive editing of kivy language code to get immediate feedback. We can see two-panel screens with a menu spinner button and other controls across the top. The left pane contains kivy code, and the right side is rendered. We can edit the left pane through changes that will be lost when we use the menu spinner button. The catalog will display dozens of .kv examples controlling different widgets and layouts. Let us see this code:

```python
import kivy
kivy.require('1.4.2')
import os
import sys
from kivy.app import App
from kivy.factory import Factory
from kivy.lang import Builder, Parser, ParserException
from kivy.properties import ObjectProperty
```

```python
from kivy.uix.boxlayout import BoxLayout
from kivy.uix.codeinput import CodeInput
from kivy.animation import Animation
from kivy.clock import Clock
CATALOG_ROOT = os.path.dirname(__file__)
# Config.set('graphics', 'width', '1024')
# Config.set('graphics', 'height', '768')
'''List of classes that need to be instantiated in the factory from .kv files.
'''
CONTAINER_KVS = os.path.join(CATALOG_ROOT, 'container_kvs')
CONTAINER_CLASSES = [c[:-3] for c in os.listdir(CONTAINER_KVS)
    if c.endswith('.kv')]
class Container(BoxLayout):
    '''A container is essentially a class that loads its root from a known
    .kv file.
    The name of the .kv file is taken from the Container's class.
    We can't just use kv rules because the class may be edited
    in the interface and reloaded by the user.
    See :meth: change_kv where this happens.
    '''

    def __init__(self, **kwargs):
        super(Container, self).__init__(**kwargs)
        self.previous_text = open(self.kv_file).read()
        parser = Parser(content=self.previous_text)
        widget = Factory.get(parser.root.name)()
        Builder._apply_rule(widget, parser.root, parser.root)
        self.add_widget(widget)
    @property
    def kv_file(self):
        '''Get the name of the kv file, a lowercase version of the class
        name.
        '''
        return os.path.join(CONTAINER_KVS, self.__class__.__name__ + '.kv')
for class_name in CONTAINER_CLASSES:
    globals()[class_name] = type(class_name, (Container,), {})
class KivyRenderTextInput(CodeInput):
    def keyboard_on_key_down(self, window, keycode, text, modifiers):
        is_osx = sys.platform == 'darwin'
        # Keycodes on OSX:
        ctrl, cmd = 64, 1024
        key, key_str = keycode
        if text and key not in (list(self.interesting_keys.keys()) + [27]):
```

```
            # This allows *either* ctrl *or* cmd, but not both.
            if modifiers == ['ctrl'] or (is_osx and modifiers == ['meta']):
                if key == ord('s'):
                    self.catalog.change_kv(True)
                    return
        return super(KivyRenderTextInput, self).keyboard_on_key_down(window,
            keycode, text, modifiers)
class Catalog(BoxLayout):
    '''Catalog of widgets. This is the root widget of the app. It contains
    a tabbed pain of widgets that can be displayed and a textbox where .kv
    language files for widgets being demoed can be edited.
    The entire interface for the Catalog is defined in kivycatalog.kv,
    although   individual   containers   are   defined   in   the   container_kvs
directory.
    To add a container to the catalog,
    first create the .kv file in container_kvs
    The name of the file (sans .kv) will be the name of the widget available
    inside the kivycatalog.kv
    Finally modify kivycatalog.kv to add an AccordionItem
    to hold the new widget.
    Follow the examples in kivycatalog.kv to ensure the item
    has an appropriate id and the class has been referenced.
    You do not need to edit any python code, just .kv language files!
    '''
    language_box = ObjectProperty()
    screen_manager = ObjectProperty()
    _change_kv_ev = None
    def __init__(self, **kwargs):
        self._previously_parsed_text = ''
        super(Catalog, self).__init__(**kwargs)
        self.show_kv(None, 'Welcome')
        self.carousel = None
    def show_kv(self, instance, value):
        '''Called when an a item is selected, we need to show the .kv language
        file associated with the newly revealed container.'''
        self.screen_manager.current = value
        child = self.screen_manager.current_screen.children[0]
        with open(child.kv_file, 'rb') as file:
            self.language_box.text = file.read().decode('utf8')
        if self._change_kv_ev is not None:
            self._change_kv_ev.cancel()
        self.change_kv()
```

```python
        # reset undo/redo history
        self.language_box.reset_undo()
    def schedule_reload(self):
        if self.auto_reload:
            txt = self.language_box.text
            child = self.screen_manager.current_screen.children[0]
            if txt == child.previous_text:
                return
            child.previous_text = txt
            if self._change_kv_ev is not None:
                self._change_kv_ev.cancel()
            if self._change_kv_ev is None:
                self._change_kv_ev = Clock.create_trigger(self.change_kv, 2)
            self._change_kv_ev()
    def change_kv(self, *largs):
        '''Called when the update button is clicked. Needs to update the
        interface for the currently active kv widget, if there is one based
        on the kv file the user entered. If there is an error in their kv
        syntax, show a nice popup.'''
        txt = self.language_box.text
        kv_container = self.screen_manager.current_screen.children[0]
        try:
            parser = Parser(content=txt)
            kv_container.clear_widgets()
            widget = Factory.get(parser.root.name)()
            Builder._apply_rule(widget, parser.root, parser.root)
            kv_container.add_widget(widget)
        except (SyntaxError, ParserException) as e:
            self.show_error(e)
        except Exception as e:
            self.show_error(e)
    def show_error(self, e):
        self.info_label.text = str(e).encode('utf-8')
        self.anim = Animation(top=190.0, opacity=1, d=2, t='in_back') +\
            Animation(top=190.0, d=3) +
            Animation(top=0, opacity=0, d=2)
        self.anim.start(self.info_label)
class KivyCatalogApp(App):
    '''The kivy App that runs the main root. All we do is build a catalog
    widget into the root.'''
    def build(self):
        return Catalog()
```

```
    def on_pause(self):
        return True
if __name__ == "__main__":
    KivyCatalogApp().run()
```

----------------------------------test.kv-----------------------------

```
#:kivy 1.4
#:import KivyLexer kivy.extras.highlight.KivyLexer
<Container>:
    canvas.before:
        Color:
            rgb: 0, 0, 0
        Rectangle:
            pos: self.pos
            size: self.size
<Catalog>:
    language_box: language_box
    screen_manager: screen_manager
    auto_reload: chkbx.active
    info_label: info_lbl
    orientation: 'vertical'
    BoxLayout:
        padding: '2sp'
        canvas:
            Color:
                rgba: 1, 1, 1, .6
            Rectangle:
                size: self.size
                pos: self.pos
        size_hint: 1, None
        height: '45sp'
        Spinner:
            size_hint: None, 1
            width: '108sp'
            text: 'Welcome'
            values: [screen.name for screen in screen_manager.screens]
            on_text: root.show_kv(*args)
        Widget:
        BoxLayout:
            size_hint: None, 1
            width: '150sp'
            Label:
```

```
                    text: "Auto Reload"
              CheckBox:
                  id: chkbx
                  active: True
                  size_hint_x: 1
        Button:
            size_hint: None, 1
            width: '108sp'
            text: 'Render Now'
            on_release: root.change_kv(*args)
    BoxLayout:
        id: reactive_layout
        orientation:  'vertical'  if  self.width  <  self.height  else
'horizontal'
        Splitter:
            id: editor_pane
            max_size: (reactive_layout.height if self.vertical else reacti-
            ve_layout.width) - self.strip_size
            min_size: sp(30) + self.strip_size
            vertical: 1 if reactive_layout.width < reactive_layout.height
else 0
            sizable_from: 'bottom' if self.vertical else 'right'
            size_hint: (1, None) if self.vertical else (None, 1)
            size: 400, 400
            on_vertical:
                mid_size = self.max_size/2
                if args[1]: self.height = mid_size
                if not args[1]: self.width = mid_size
            ScrollView:
                id: kr_scroll
                KivyRenderTextInput:
                    catalog: root
                    id: language_box
                    auto_indent: True
                    lexer: KivyLexer()
                    size_hint: 1, None
                    height: max(kr_scroll.height, self.minimum_height)
                    valign: "top"
                    text: "This box will display the kivy language for what-
                    ever has been selected"
                    on_text: root.schedule_reload()
                    on_cursor: root.schedule_reload()
```

```
ScreenManager:
    id: screen_manager
    Screen:
        name: "Welcome"
        PlaygroundContainer:
    Screen:
        name: "Float Layout"
        FloatLayoutContainer
    Screen:
        name: "Box Layout"
        BoxLayoutContainer:
    Screen:
        name: "Anchor Layout"
        AnchorLayoutContainer:
    Screen:
        name: "Grid Layout"
        GridLayoutContainer:
    Screen:
        name: "Stack Layout"
        StackLayoutContainer:
    Screen:
        name: "Buttons"
        ButtonContainer:
    Screen:
        name: "Labels"
        LabelContainer:
    Screen:
        name: "Booleans"
        CheckBoxContainer:
    Screen:
        name: "Progress Bar"
        ProgressBarContainer:
    Screen:
        name: "Media"
        MediaContainer:
    Screen:
        name: "Text"
        TextContainer:
    Screen:
        name: "Popups"
        PopupContainer:
```

```
            Screen:
                name: "Selectors"
                SelectorsContainer:
            Screen:
                name: "File Choosers"
                FileChooserContainer:
            Screen:
                name: "Scatter"
                ScatterContainer:
            Screen:
                name: "ReST"
                RestContainer:
FloatLayout:
    size_hint: 1, None
    height: 0
    TextInput:
        id:info_lbl
        readonly: True
        font_size: '14sp'
        background_color: (0, 0, 0, 1)
        foreground_color: (1, 1, 1, 1)
        opacity:0
        size_hint: 1, None
        text_size: self.size
        height: '150pt'
        top: 0
```

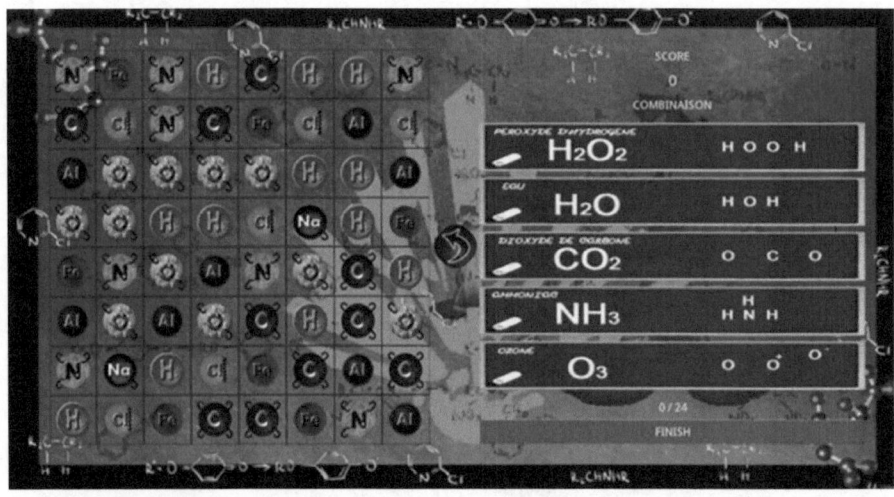

Figure 13.5: Video game.

13.5 Creating my own screen

We can design our own screen by just extending screen class as given in the code:

```python
from kivy.app import Apps
from kivy.graphics.vertex_instructions import Rectangle
from kivy.lang import Builder
from kivy.uix.screenmanager import Screen, ScreenManager
class MyScreen(Screen):
    def __init__(self):
        super(MyScreen, self).__init__()
        # make the Rectangle here and save a reference to it
        with self.canvas.before:
            self.rect = Rectangle(source='two.png')
    def on_pos(self, *args):
        # update Rectangle position when position changes
        self.rect.pos = self.pos
    def on_size(self, *args):
        # update Rectangle size when size changes
        self.rect.size = self.size
    def CheckEvent(self):
        if self.ids.door1.source == "Clicked":
            print("Button has been clicked")
            # use the saved reference to change the background
            self.rect.source = 'one.png'
        else:
            print("no correct button")
Builder.load_string('''
<MyScreen>:
    FloatLayout:
        Button:
            pos_hint: {"top": .8, "right": .75}
            size_hint: .5, .1
            text:
                "Click Me"
            source: "Clicked"
            id: door1
            on_press:
                root.CheckEvent()
''')
class TestApp(App):
    def build(self):
```

```
        sm = ScreenManager()
        sm.add_widget(MyScreen())
        return sm
TestApp().run()
```

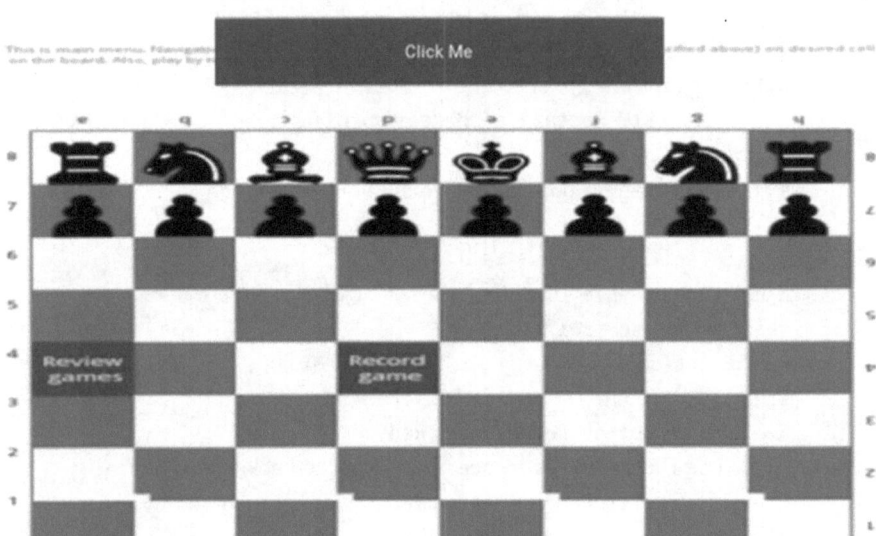

Figure 13.6: Chess game.

13.6 Project game

In this project, we are going to design a game. Let us see how to do it:

```
from kivy.app import App
from kivy.uix.button import Button
from kivy.uix.screenmanager import Screen
from kivy.uix.widget import Widget
class MenuButton(Button):
    pass
class GameScreen(Screen):
    def __init__(self, **kwargs):
```

```
          super(GameScreen, self).__init__(**kwargs)
class CellButton(Button):
    def __init__(self, row, column):
        super(CellButton, self).__init__()
        self.text = 'HI'
        self.color = 'BLUE'
class Game(Widget):
    # grid = ObjectProperty(None)
    def __init__(self, **kwargs):
        super(Game, self).__init__()
        self.difficulty = 'GAME_DIFFICULTY'
        self.InitializeGrid
        self.ids.grid.add_widget(CellButton(1, 2))
    def InitializeGrid(self):
        for i in range(6):
            self.ids.grid.add_widget(CellButton(1, 2))
    def rebuild_background(self):
        self.canvas.before.clear()
class MyGameApp(App):
    '''The kivy App that runs the main root. All we do is build a catalog
       widget into the root.'''
    def build(self):
        return Game()
if __name__ == "__main__":
    MyGameApp().run()

----------------------------------mygame.kv----------------------------
<GameScreen>:
BoxLayout:
  orientation: 'vertical'
  canvas:
    Color:
      rgb: 1,0,1,1
    Rectangle:
      pos: self.pos
      size: self.size
  Label:
    text: 'Totally better than FlappyBird. Will sell out for millions'
    color: 0,1,0,1
    width: root.width
    valign: 'top'
    height: min(root.height, root.width) / 6.
```

```
      font_name: 'Arial'
      font_size: root.height / 12
Game:
   size_hint: None, None
   size: root.width, root.width
BoxLayout:
   size: root.width, root.height
   MenuButton:
      text: 'button1'
   MenuButton:
      text: 'button2'
   MenuButton:
      text: 'button3'
   MenuButton:
      text: 'button4'
   MenuButton:
      text: 'button5'
   MenuButton:
      text: 'button6'
<Game>:
   GridLayout:
      id: grid
      pos:root.pos
      size: root.size
      cols: 6
      rows: 6
      canvas:
         Color:
            rgb: 1,0,0,1
         Rectangle:
            pos: self.pos
            size: self.size
<MenuButton>:
   GridLayout:
      id: grid
      pos:root.pos
      size: root.size
      cols: 6
      rows: 6
      canvas:
```

```
Color:
  rgb: 1,0,0,1
Rectangle:
  pos: self.pos
  size: self.size
```

Figure 13.7: Game.

Bibliography

www.tutorialspoint.com/kivy
www.developer.android.com/training
www.javatpoint.com/android-tutorials
www.vogella.com/tutorials/android/article.html
https://kivy.org/doc/stable/
https://docs.python.org/3/
https://stackoverflow.com
https://developer.android.com/docs
https://likegeeks.com/kivy-tutorial/
https://riptutorial.com/kivy
https://www.codementor.io/kiok46/beginner-kivy-tutorial-basic-crash-course-for-apps-in-kivy
    -y2ubiq0gz
https://pythonprogramming.net/kivy-application-development-tutorial/
https://towardsdatascience.com/python-for-android-start-building-kivy-cross-platform-
    applications-6cf867d44612
https://techwithtim.net/tutorials/kivy-tutorial/setup/
https://github.com

https://doi.org/10.1515/9783110689488-014

Index

https://doi.org/10.1515/9783110689488-015